Preventing Industrial Toxic Hazards

A Guide for Communities

Revised edition of *A Citizen's Guide to Promoting Toxic Waste Reduction*

Marian Wise and Lauren Kenworthy

INFORM, Inc.
381 Park Avenue South
New York, NY 10016-8806
Tel   (212) 689-4040
Fax   (212) 447-0689

Library of Congress Cataloging-in-Publication Data

Wise, Marian, 1957 -
      Preventing industrial toxic hazards : a guide for communities /
Marian Wise and Lauren Kenworthy. -- 2nd ed.
            p.   cm.
      Rev. ed. of: A citizen's guide to promoting toxic waste reduction. 1990.
      Includes bibliographical references.
      ISBN 0-918780-60-8 : $25.00

      1. Hazardous wastes--Management--Citizen participation.
2. Factory and trade waste--Management--Citizen participation.
3. Waste minimization--Citizen participation. I. Kenworthy,
Lauren, 1962-   . II. Kenworthy, Lauren, 1962-   Citizen's guide to
promoting toxic waste reduction. III. Title.
TC1050.C57K47   1993
363.72 ' 87525--dc20   92-41873
                                                CIP

INFORM, Inc., founded in 1974, is a nonprofit research organization that identifies and reports
on practical actions for the protection and conservation of natural resources and public
health. INFORM's research is published in books, articles, and a quarterly newsletter. Its
work is supported by contributions from individuals and corporations and by grants from
foundations and the Federal government.

Cover design by Saul Lambert

Printed on recycled paper

Table of Contents

Preface

Since its publication in 1990, INFORM's *A Citizen's Guide to Promoting Toxic Waste Reduction* has been used and praised by community groups and by many companies as well across the United States. With it, concerned citizens have learned how to use the US Environmental Protection Agency's Toxics Release Inventory and other data to understand the operations and discharges of local plants, and they have had guidance in building positive communications with plant officials and for gathering the information they need to assess a plant's source reduction efforts. We have been gratified to see the guide's value demonstrated repeatedly, producing constructive dialogue between grassroots organizations and local industries once perceived as polluting adversaries.

Over the past decade INFORM's Chemical Hazards Prevention Program has produced 13 published works that have documented the significant environmental and economic benefits of pollution prevention and stimulated progress at individual plants and in the chemical industry at large. Our research has highlighted successful existing industrial efforts and identified public-policy options for promoting further progress. INFORM outreach has followed our research, making our findings useful to public and private sector leaders alike.

The need for insights on preventive strategies for addressing US toxic problems is great. Public concerns about toxics are rising worldwide, both for the environmental and health hazards they pose and for the skyrocketing costs of control efforts: By 1990, treatment and disposal costs in the United States exceeded $100 billion a year — an insupportable economic toll on industry and government, and one that did not even achieve its environmental goal of assuring the protection of our natural resources and the health of the people.

INFORM is proud to present *Preventing Industrial Toxic Hazards: A Guide for Communities,* the first comprehensive update of our *Citizen's Guide.* It contains information on how to take advantage of many more resources, such as the RCRA Hazardous Waste Reports, occupational health and safety information, data on accidents and spills, and other data. We've redesigned our worksheets to make them easier to use; revised and annotated our list of published and electronic resources, and much more.

We hope citizens' groups and industry alike will find this guide an even more valuable resource. We hope you will use it in good health!

Joanna D. Underwood
President, INFORM

Acknowledgments

We wish to thank several people for their valuable contributions to this guide. Many creative suggestions were offered by the manuscript's reviewers: Warren R. Muir of Hampshire Research Associates, who is also INFORM's science director; Rick Engler, vice president of the New Jersey Industrial Union Council, AFL-CIO; Ed Hopkins and Sandy Buchanan of Ohio Citizen Action; Christopher J. Daggett, managing director of William E. Simon & Sons, Inc. (formerly the Commissioner of the New Jersey Department of Environmental Protection); and Eric Schaeffer, Director of the Pollution Prevention Policy Staff at the US Environmental Protection Agency.

Thanks also to the INFORM staff members who contributed to the successful completion of this project: Mark H. Dorfman, senior associate in INFORM's Chemical Hazards Prevention program, for his technical assistance on source reduction; Joanna D. Underwood, president of INFORM, for her support throughout research, writing, and production of the book; Sibyl R. Golden, former director of research and publications, who helped make the guide more readable and usable to concerned citizens; and Joseph E. Mohbat, director of research and publications, for guiding the manuscript through the final editing process and seeing it through to completion.

Many thanks also to Sharene Azimi, assistant editor, for meticulous copy editing and proofreading; to Elisa Last, production coordinator, for creative design and production; to Michèle Ascione, research assistant, for her help with miscellaneous research and typing of the manuscript; and to Flora Che, an INFORM intern, for her help in double-checking the information sources.

Finally, INFORM thanks the many individuals and organizations that provided general support, the US Environmental Protection Agency, and the following foundations whose grants made possible this revision of the original *Citizen's Guide:* The Educational Foundation of America, The Joyce Foundation, the Geraldine R. Dodge Foundation, the Victoria Foundation, and the James C. Penney Foundation. EPA's support does not signify that the contents necessarily reflect the views and policies of EPA.

While the information in this book could not have been collected without the assistance of all these people, the findings and conclusions are the sole responsibility of INFORM.

Chapter 1 Introduction

Since World War II, synthetic chemicals have played an increasingly large role in our daily lives. Almost without becoming aware of it, we have grown dependent on medicines, plastics, automobiles, synthetic textiles, and myriad other products to maintain our standard of living. Chemistry has enabled us to live more comfortably, surrounded by many conveniences that we take for granted.

"Better living through chemistry" has come with a price, however. In 1990, the United States Environmental Protection Agency (EPA) reported almost 5 billion pounds of toxic chemicals discharged into our air and water, injected underground, or carried off industrial sites for treatment, storage, and disposal. *(See Box 1-1: Toxic Chemicals in Our Environment.)* Toxic chemicals reported include known carcinogens such as benzene, dichloromethane, and styrene. This 5 billion pounds represents the wastes of only 302 types of chemicals and 20 chemical groups, just a fraction of the more than 70,000 chemicals in commercial use today. Further, only large facilities within the manufacturing sector report their emissions to EPA. This neglects other potentially large sources of toxic releases, such as those in power plants, federal facilities, filling stations, dry cleaners, and small businesses.

According to EPA estimates, 1,000 new chemicals are proposed for manufacture each year.[1] In just 40 years, annual production of synthetic organic chemicals has increased 15-fold in volume, changing our way of life.[2]

Figures such as these have captured the attention of a growing number of citizens who are concerned about the potential damage that **toxic or hazardous** substances may cause to workers and communities.

The Need for Reducing Toxic Chemical Wastes at the Source

The Federal government and state governments have responded to the need for reducing toxic chemical wastes at the source by enacting a growing number of environmental laws. These laws have been strict in some areas (e.g., outright bans on the use of certain chemicals), but much less strict in others (e.g., air and wa-

Box 1-1: Toxic Chemicals in Our Environment

The federal Emergency Planning and Community Right-to-Know law, enacted in 1986, created a mechanism by which large industrial facilities must publicly report their annual emissions of toxic chemicals into the environment. This information is collected by EPA and made available to the public as the **Toxics Release Inventory (TRI).**

In 1990, the most recent year for which such figures are available, EPA received more than 83,387 Toxics Release Inventory reports from 23,638 facilities that use or manufacture chemicals. According to the 1990 TRI data, 4.83 billion pounds of toxic chemicals were released or transferred off-site for treatment or disposal: 2.20 billion pounds of toxic chemicals were discharged into the air; 197 million pounds were released to bodies of water; 441 million pounds were released to land; 725 million pounds were injected underground; an estimated 448 million pounds were transferred to wastewater treatment plants for processing and disposal; and 815 million pounds were sent to other off-site locations for treatment, storage, or disposal.[3]

ter permits that provide a legal right to discharge many types of toxic or hazardous substances into the environment). However, it has become clear that, for several reasons, pollution control regulations alone cannot protect human health and the environment:

- Many chemicals remain unstudied, unmonitored, and unregulated. Even in the area of land disposal, which has received intensive attention from regulators, the US General Accounting Office found that the Environmental Protection Agency "does not know whether it is controlling 90 percent of existing hazardous waste or 10 percent."[4]

- Many regulations are technically complex and can take years, or even decades, to implement. Some become the subject of protracted legal battles, lengthening the process still further and making the outcome uncertain. For example, after EPA worked more than 15 years on a rule that would have required a phasing out of the use of asbestos over 10 years, a Federal court overturned the entire effort.[5]

- Officials charged with granting environmental permits and enforcing environmental and occupational regulations are limited in their activities on the federal and state levels. Lean budgets limit funds for their activities, and large budget deficits foreclose hopes for the major spending increases needed to turn this situation around.

- A recent study conducted by a former researcher for the congressional Office of Technology Assessment (OTA) predicts a 75 percent increase in the generation of hazardous waste in the United States from 1988 to 2000.[6]

Just as regulators are being overwhelmed by the number of pollution problems facing this country, the public is coming to understand that no matter how many good pollution control laws are passed, the planet is simply too small to handle the volumes of toxic or hazardous wastes dumped onto it each year. Pollution control systems have often ended up simply moving pollutants from one environmental medium to another, rather than actually getting rid of them. Incinerators, for example, consume toxic or hazardous wastes from land dumps, often only to turn some of the toxic constituents into air pollutants, while smokestack scrubbers that capture air pollutants often create solid waste in the form of ash. This often creates a "toxics shell game," which is difficult, if not impossible, to win.

At the same time, industry has realized that the liability for cleaning up old pollutants, plus the cost of managing new pollution, add up to an enormously expensive problem. Of the more than $100 billion spent on federally mandated pollution control and clean-up programs in 1990, business paid two-thirds of the bill; government paid the rest. [7]

Other Toxics Exposures

Besides widespread environmental contamination and costly pollution control, toxic chemical use has posed other risks: risks to plant workers, and risks to public health when these chemicals are spilled during storage or transport. Greater awareness of these potential hazards has spawned heightened concern among environmental, community, and labor organizations and individuals, all of whom are increasingly demanding better protection and broad-based solutions.

- **Occupational exposures** — About 32 million workers face potential exposure to one or more of an estimated 575,000 existing chemicals and chemical mixtures.[8] According to the Workplace Health Fund, medical experts estimate that some 10 million people in the United States are at high risk of developing a serious illness as a result of past or present exposure to toxic substances in the workplace.

  In a 1989 study on occupational disease in New Jersey, conducted at Mt. Sinai Medical School in New York, the federal Occupational Safety and Health Administration repeatedly found New Jersey workers over-exposed (exposed to substances in quantities above the "permissible exposure limit" as designated under the Occupational Safety and Health Act) to lead, silica, and several known carcinogens, such as asbestos, arsenic, and nickel. Some worksites in the state were found to present potentially serious exposure to other toxic materials, such as formaldehyde, benzene, and methylene chloride: formaldehyde and benzene are also known human carcinogens. The report maintained that environmental contamination at hazardous waste sites in New Jersey is widespread. Such contamination, according to the study, may present a potential health hazard to community residents as well as to workers in the hazardous waste industry.[9]

- **Unauthorized or accidental releases and spills involving hazardous substances** — These are regulated under Section 9601 (14) of the Comprehen-

Box 1-2: "Toxic or Hazardous" Substances and Wastes

The term "toxic or hazardous," often used to describe certain types of chemicals and wastes, is easily subject to misinterpretation because the words mean different things in different contexts. INFORM's use of this term is broader than conventional uses, which often refer only to certain materials on chemical lists created by environmental laws and regulated by the Federal government. For example, "toxic or hazardous" substances appear on lists in several environmental laws, such as the **Clean Air Act** (hazardous air emissions), **Federal Water Pollution Control Act** (priority pollutants), **Resource Conservation and Recovery Act** (hazardous constituents, hazardous wastes, acute hazardous wastes), **Comprehensive Environmental Response, Compensation and Liability Act** (hazardous substances), and **Emergency Planning and Community Right-to-Know Act** (toxic chemicals, extremely hazardous substances).

Toxic or hazardous substances are defined somewhat differently in each of these laws. For instance, a chemical can be added to the Toxics Release Inventory list as a "toxic chemical" under the Emergency Planning and Community Right-to-Know Act (EPCRA) if the Administrator of EPA determines that there is sufficient evidence to establish any one of the following criteria: the chemical is known to cause or may cause significant adverse human health effects when released beyond facility boundaries; the chemical is a known or suspected carcinogen or may cause reproductive, neurological, or heritable disorders, or other chronic health effects; or the chemical's toxicity may persist or bioaccumulate in the environment. (EPCRA, Section 313(d)(2).) At the same time, the Comprehensive Emergency Response, Compensation and Liability Act (CERCLA) defines "hazardous substance" to include substances covered in a number of existing environmental laws, as well as any "element, compound, mixture, solution or substance which, when released into the environment, may present substantial danger to the public health or welfare or the environment...." (CERCLA, Sections 9601(14) and 9602(a)).

---

sive Environmental Response, Compensation and Liability Act (CERCLA). Between 1987 and 1991, 85,000 non-routine oil discharges and more than 28,000 incidents involving hazardous substances were reported to the National Response Center (NRC) and EPA (see Chapter 5).[10]

Reportable accidents and spills continue to occur. For example, in late June 1992, a freight train derailment caused the rupture of a tank car containing benzene, sending a vapor cloud over sections of Minnesota and Wisconsin and contaminating a river that flows into Lake Superior. At least 50,000 people evacuated the area, many complaining of dizziness, headaches, and burning eyes and skin. Benzene, which can cause leukemia in humans, is a flammable liquid used as a solvent and as a raw material in making plastics, insecticides, detergents, and paints.

Source reduction is the one strategy that simultaneously addresses the problems of environmental contamination, occupational exposures, and accidents and spills, for this strategy alone focuses on not producing a waste in the first place. Source reduction means increasing the efficient use of toxic or hazardous substances and reducing or eliminating the generation of hazardous or toxic wastes, instead of dealing with the problems after the fact. It is a strategy that protects hu-

For the purpose of this guide, INFORM uses the term "toxic or hazardous" to include not only substances regulated under federal laws, but any of the more than 70,000 known chemical substances that may not be regulated but may, nonetheless, fall within the federal criteria for defining toxicity. For example, various chemicals listed as known or probable carcinogens by the US Department of Health and Human Services (HHS) and the International Agency for Research on Cancer (IARC) are not covered by Toxics Release Inventory reporting requirements. However, the appearance of the chemical on these HHS and IARC lists suggests that it is a suspected, if not confirmed, carcinogen, and that it is worth tracking.

Toxic or hazardous waste often refers to materials that companies must report to the Toxics Release Inventory and to solid hazardous wastes regulated under the Resource Conservation and Recovery Act (RCRA). INFORM considers any type of wastestream hazardous or toxic if it contains one of the substances appearing on any of the lists of substances regulated by the federal or state governments, whether or not the substance is regulated in the medium in which it is released, and regardless of its concentration.

The term "toxic or hazardous waste" as used in this guide is not intended to imply that the concentration and amount of the chemicals in the various wastestreams are always present in quantities sufficient to cause harm. The focus is essentially on chemicals that are generated as waste, or may potentially result in occupational exposures or accidents, and does not include an analysis of the extent or severity of environmental or human exposures to these substances.

man health and the environment, while saving money. Through source reduction, companies can operate more efficiently — wasting fewer raw materials, creating less pollution, and reducing their pollution control, handling, and liability costs.

Promoting Source Reduction: The Role You Can Play

Citizens can play a major role in making source reduction happen. You can have a strong and positive impact by challenging your local plant to be a good neighbor and work with surrounding communities to remedy those toxics problems that environmental laws and regulations have not yet solved. Source reduction is often the ideal tool for this task. But source reduction is still a relatively new idea for many in industry, and businesses may not fully explore source reduction opportunities without the urging of people like you. Even in states where source reduction programs already are in place, there may be some difficulty with industry cooperation or inadequate enforcement of the program within the responsible government agency. That is why your involvement is so important.

How This Guide Will Help You

This guide will help you study and evaluate the source reduction efforts of factories in your community and source reduction programs in your state. With it, you can actively promote source reduction and encourage the reduction or elimination of potentially dangerous chemical discharges, whether or not they are currently regulated by law. The guide equips you with the facts you need to show businesses in your neighborhood why source reduction, as the environmental and human health protection method of choice, should be the first line of defense against pollution. By following the steps for researching and evaluating local plants, you will be able to:

- **Conduct detailed background research** on a plant.

- **Engage in a cooperative effort with management and workers** at a plant and enlist their support in promoting source reduction.

- **Create a dialogue** with company representatives and spur the company to explore source reduction opportunities.

- **Assess whether a company has created the policies,** management incentives, and informational tools required for a serious source reduction program.

- **Identify specific source reduction actions** taken at a company and draw basic conclusions about the success of the company program over time.

This guide will not enable you to evaluate the technical aspects of specific source reduction actions taken, because this requires detailed knowledge and understanding of the individual production processes involved. However, it will allow you to judge the overall source reduction program and send a powerful message that you are expecting responsible, documentable initiatives from local companies when it comes to preventing pollution and other chemical hazards through source reduction. In the long run, this guide can help you help local businesses to become better, cleaner neighbors.

Contents of the Chapters

Chapters 2-5 provide you with the tools needed to promote source reduction at individual plants in your community. Each chapter steers you to experts and publications that provide detailed facts and different perspectives on source reduction. A list of key sources and important documents follows each chapter.

Chapter 2 introduces you to the concept of source reduction, describes its benefits, and explains how it is fundamentally different from waste management and pollution control methods.

Chapters 3 and 4 describe the strategies plants can use to reduce toxic or hazardous substances at the plant, and the steps company management needs to take to create effective source reduction programs.

Chapter 5 gives you a step-by-step research process to follow when studying local plants, including worksheets and questions for interviews with company representatives.

Chapter 6, to help you promote source reduction, outlines some of the legislative approaches that can support your efforts at the local level by enabling you to get more information on local plants and their pollution records and by encouraging industrial source reduction through new incentives and requirements.

The **Appendices** provide information on state and federal contacts, tables on the health effects of Toxics Release Inventory chemicals, and sample forms used by plant officials for public reporting.

Although this guide has been written for readers who do not have a background in industrial chemistry or other technical areas, it has sometimes been necessary to introduce a few technical terms. These terms are highlighted with boldface when they are first used, defined in the text, and listed in the Glossary (Appendix G).

Notes

1. US Environmental Protection Agency Office of Pollution Prevention and Toxics (TS-779), *1990 Toxics Release Inventory, Public Data Release,* EPA 700-S-92-002, May 1992, p. 4.

2. US Environmental Protection Agency Office of Toxic Substances, Toxic Substances Control Act Hotline (phone communication with Marian Wise).

3. Worldwatch Institute, *Defusing the Toxics Threat: Controlling Pesticides and Industrial Waste,* Worldwatch Paper No. 79, 1987.

4. US General Accounting Office, *New Approaches Needed to Manage RCRA,* RCED-88-115, 1988, p. 19.

5. *Corrosion Proof Fitting vs. EPA,* 947 F.2d. 1201 (5th Cir. 1991)

6. Joel S. Hirschorn and Kirsten U. Oldenburg, *Prosperity Without Pollution, The Prevention Strategy for Industry and Consumers,* New York: Van Nostrand Reinhold, 1991, p. 119.

7. US Environmental Protection Agency, cited in *The New York Times,* December 23, 1990.

8. US Department of Labor, Occupational Safety and Health Administration, *Chemical Hazards Communication,* 1988.

9. Division of Environmental and Occupational Medicine, Department of Community Medicine, Mt. Sinai School of Medicine, *Occupational Disease in New Jersey,* December 1989, p. 21.

10. US Environmental Protection Agency Office of Solid Waste and Emergency Response, *An Overview of ERNS, Emergency Response Notification System Fact Sheet,* Publication 9360.0-29FS, April 1992.

Chapter 2 The Source Reduction Option

The term "source reduction" has been used in a variety of ways. It is important that you and the industrial representatives you deal with have a shared understanding of the concept: both what it includes and what it does not include. Thus, it is extremely important to understand the essential characteristics of source reduction.

What is Source Reduction?

The passage of the federal **Pollution Prevention Act of 1990** marked the emergence of source reduction as national policy. The preamble of the new law, illustrating congressional recognition of the benefits of source reduction over waste treatment and disposal, states:

> "There are significant opportunities for industry to reduce or prevent pollution at the source through cost-effective changes in production, operation, and raw materials use. Such changes offer industry substantial savings in reduced raw material, pollution control, and liability costs as well as help protect the environment and reduce risks to worker health and safety."

Section 6603 of the Pollution Prevention Act defines "source reduction" as "any practice that: (i) reduces the amount of any hazardous substance, pollutant, or contaminant entering any waste stream or otherwise released into the environment (including fugitive emissions) prior to recycling, treatment, or disposal; and (ii) reduces the hazards to public health and the environment associated with the release of such substances, pollutants or contaminants."

The definition of "source reduction" in the federal law also lists the five source reduction methods:

1. Equipment or technology modifications: also called **equipment changes,** meaning modifications of and additions to equipment used in any stage of

the manufacturing process (e.g., equipment used for storing, moving, or re-acting chemicals) in order to reduce the amount of waste generated.

2. Process or procedure modifications: also called **process changes,** meaning any change in the production process that reduces the generation of waste, ranging from simple alterations of process conditions, such as temperature and pressure, to discovery of new chemical pathways and production technologies.

3. Reformulation or redesign of products: also called **product changes,** meaning changes in the product itself that can be achieved without changing the fundamental manufacturing process and that reduce the initial use of the toxic chemical or the generation of waste (e.g., creating a chemical product in the form of pellets rather than as a powder to reduce the generation of waste dusts as the material is transferred during final packaging operations).

4. Substitution of raw materials: also called **chemical substitution,** meaning replacement of toxic or hazardous chemicals with nonhazardous or less hazardous chemicals in both production and nonproduction processes.

5. Improvements in housekeeping, maintenance, training, or inventory control: also called **operational changes,** meaning changes in the way hazardous materials are handled at a plant (e.g., careful observation and control of materials, process conditions, and employee habits in order to minimize spills, process upsets, or use of excessive amounts of chemicals) that can reduce the generation of waste.

The law also explicitly states that "source reduction does not include any practice which alters the physical, chemical, or biological characteristics or the volume of a hazardous substance, pollutant, or contaminant through a process or activity which itself is not integral to any process necessary for the production of a product or the providing of a service." In other words, the law states that source reduction does not include practices that involve the management of wastes already created. In fact, both the preamble of the Pollution Prevention Act and the definition of source reduction focus upon the reduction of hazardous materials and pollutants before they enter the waste stream.

INFORM and the congressional Office of Technology Assessment (OTA) define source reduction as a strategy for reducing pollution that involves preventing the generation of waste in the first place rather than cleaning it up, treating it, or recycling after it has been produced.

The federal law, INFORM, and OTA limit their definitions of source reduction to action taken prior to **out-of-process recycling (recycling), treatment, disposal,** or any other form of managing waste. All three present source reduction options — some technical and some not — that either increase the efficiency of chemical use in a plant or eliminate the use of a chemical altogether. Although you may hear plant managers or agency officials talk about "waste minimization" or "pollution prevention" in the context of source reduction, these terms generally are not synonymous *(See Box 2-1: Why "Waste Minimization" and "Pollution*

Prevention" Are Not Necessarily "Source Reduction.") This guide uses the term "source reduction" because it is the term most consistently used to refer to the reduction of toxic or hazardous materials and the generation of wastes prior to any waste management practices.

Why Is Source Reduction the Best Strategy?

Source reduction is preventive. By reducing or eliminating toxics and the generation of waste at the source, source reduction prevents a variety of hazards. Unlike **end-of-the-pipe** pollution control strategies, which address the problem after it has been created, source reduction can make managing, handling, storing, or transporting a waste unnecessary. It can reduce the amount of pollutants transported from, spilled at, or emitted from a factory and prevent the transfer of pollution from one part of the environment to another: the transfer that occurs, for example, when sludge from air pollution control scrubbers is dumped on the land, or toxic chemicals from solid waste seep into groundwater supplies.

Source reduction protects workers, communities, and the environment better than any pollution control technique can. Therefore, it is considered the optimal choice by the United Nations, Congress, the US Environmental Protection Agency (EPA), the Chemical Manufacturers Association, Greenpeace International, and many other organizations.

It is important to remember, however, that even though a plant may have an excellent source reduction program, there may still be hazardous waste generated that must be managed in the most environmentally sound manner possible. The federal Pollution Prevention Act of 1990 addresses the prioritization of waste management techniques as follows: "Congress hereby declares it to be the national policy of the United States that pollution should be prevented or reduced at the source, whenever feasible; pollution that cannot be prevented should be recycled in an environmentally safe manner, whenever feasible; pollution that cannot be prevented or recycled should be treated in an environmentally safe manner; and disposal or other release to the environment should be employed only as a last resort and should be conducted in an environmentally safe manner."

Thus, in keeping with the federal law, EPA ranks various waste management techniques below source reduction in an "environmental protection hierarchy." *(See Box 2-2: Environmental Protection Hierarchy.)* Recycling is considered the next best way to handle waste. Treatment — such as biological degradation of toxic chemicals, well-controlled incineration of hazardous waste, or smokestack scrubbers that remove pollutants from air emissions — is next in line. Legally permitted disposal (for example, putting waste-filled barrels in landfills) is the least desirable waste management technique.

Source reduction prevents all environmental pollution. Source reduction is called a **multimedia** approach because it addresses pollution going to all parts, or media, of the environment: land, water, and air. *(See Box 2-3: Source Reduction Applies to All Toxic Pollutants.)* Candidates for source reduction include plant air emissions, whether intentional **point-source emissions** (air pollutants released through smokestacks, vents, and other intentional release points) or unintentional **fugitive emissions** (air pollutants released through leaky valves, evaporation from tanks, and other unintentional release points); wastewater discharges; and releases of solid wastes, whether they are sent to incinerators, **publicly owned treatment works** (sewage treatment facilities), other forms of treatment, or landfills. Source reduction strategies can be used regardless of whether plant releases are currently regulated by state or federal laws. Source reduction avoids the so-called "toxics shell game" since a pollutant not created is not there to be moved around.

Source reduction can prevent health hazards. EPA has documented that pollutants can pose serious health hazards in all parts of the environment. For example, discharges of toxic chemicals into the air, which have received the least government and industry attention for many years, may pose some of the most serious health threats of all. One hundred eighty-nine of these air toxics were added to the list of toxic air pollutants in the Clean Air Act Amendments of 1990.

Source reduction is often cost-effective. Pollution management strategies are expensive: as a nation we spent more than $100 billion in 1990 on federally mandated pollution control and cleanup. Source reduction, on the other hand, often saves industry money by avoiding loss of valuable raw materials and end products that were formerly discharged into the environment and by reducing pollution management costs arising from waste handling, regulatory compliance, and liability. For example, under the 1976 RCRA regulations, any site storing "hazardous" chemicals had to be classified as a hazardous waste site. As a result, the American Cyanamid plant in Marietta, Ohio, which used the hazardous chemical nitrobenzene, responded to high handling and monitoring costs by eliminating the use of nitrobenzene as a solvent for the production of a yellow dye. The company now saves $200,000 a year — mainly through reduced disposal costs, but also through avoiding the energy costs that had been necessary to recover nitrobenzene.

Box 2-3: Source Reduction Applies to All Toxic Pollutants

Over the years, the Federal government has developed several lists of toxic or hazardous pollutants, each relating to a different law. Some of these laws group pollutants by the medium – land, air, or water – that is the subject of the regulations. Others stipulate procedures plants must follow if they use certain chemicals. Source reduction targets the chemicals on all of these lists.

Clean Air Act Hazardous Air Emissions. A list of 189 toxic air pollutants and **criteria pollutants** (six common air pollutants – sulfur dioxide, carbon monoxide, particulates, nitrogen oxides, ozone and lead – for which EPA set national ambient air quality standards [NAAQS] before 1990 under Section 108 of the Clean Air Act).

Clean Water Act Priority Pollutants. A list of 126 individual chemicals released into the water, including volatile organic substances such as benzene, chloroform, and vinyl chloride; acid compounds such as phenol and its derivatives; pesticides such as chlordane, DDT, and toxaphene; heavy metals such as lead and mercury; PCBs; and other organic and inorganic compounds.

Resource Conservation and Recovery Act (RCRA) Hazardous Wastes. More than 400 discarded commercial chemical products and specific chemical constituents of industrial waste streams destined for treatment or disposal on land.

Superfund Amendments and Reauthorization Act (SARA) Title III Section 313: Toxic Substances. More than 300 chemicals and chemical categories released into all three environmental media; under specified conditions, facilities must report releases of these chemicals to EPA's annual Toxics Release Inventory.

Superfund Amendments and Reauthorization Act (SARA) Section 302: Extremely Hazardous Substances. More than 360 chemicals for which facilities are required to prepare emergency action plans if these chemicals are present in quantities above certain threshold levels. Section 304 of SARA requires facilities to report releases of these chemicals to the air, land, or water to the State Emergency Response Committee (SERC) and the Local Emergency Planning Committee (LEPC).

Source reduction measures also may be applied to other potentially toxic or hazardous substances not yet singled out for regulation, or to the balance of the 70,000 chemicals in commercial use today, whether or not their toxicology is fully understood.

Source reduction offers "win-win" solutions to environmental problems by accomplishing better environmental protection and saving industry money at the same time. A single Dow Chemical plant, for example, saves more than $2.4 million dollars per year through just one source reduction technique that took only about four months to implement.

Source reduction is often easy to carry out. Although some source reduction techniques can be difficult to identify and implement, most include many low- or no-technology approaches that require little capital to implement. For example, a simple housekeeping change at the CIBA-GEIGY plant in New Jersey involved the handling of quality control samples. Before 1985, product samples were analyzed and then disposed of, for fear that if a sample was returned to the wrong batch, the whole batch might become waste. Since 1985, the plant has taken the minimum sample size needed and returned the samples to the process. This source reduction technique has reduced the wastes created by quality control sampling by 50 percent to 20,000 pounds; it saves the company $100,000 a year.

Source reduction has tremendous potential. *Environmental Dividends: Cutting More Chemical Wastes,* INFORM's 1992 study of source reduction practices at 29 organic chemical plants, revealed impressive source reduction results, even though complete information on waste and cost savings was not available for all 181 source reduction activities documented. For example, just 80 of the activities resulted in annual reductions of 128.7 million pounds of toxic or hazardous wastes. Similarly, just 62 of the activities resulted in annual savings of $21.8 million.

Overall, the INFORM study plants dramatically decreased their releases of toxic or hazardous wastes, saved money, and increased product yields through source reduction. Yet these plants represent only a fraction of the waste-generating facilities in the United States and around the world. Based on the growing concerns about these wastes, and on the specific findings documented in *Environmental Dividends,* INFORM has drawn several conclusions:

1. Source reduction offers waste-generating facilities continuing opportunities to reduce significantly the amounts of toxic or hazardous wastes they generate and subsequently release into the air, land, or water or transfer to treatment or disposal facilities. Even plants that have been implementing source reduction for many years continue to achieve large reductions.

2. Source reduction also offers waste-generating facilities continuing opportunities to reduce costs rapidly and increase production efficiency:

    - The cost of implementing many source reduction activities is low: one-quarter of the source reduction activities in INFORM's study for which capital cost data were provided required no capital investment, and nearly half of the source reduction activities required investments of less than $100,000.

    - Companies can recoup their initial investments rapidly: in six months or less for two-thirds of the source reduction activities in INFORM's study for which information on payback periods was reported.

    - Implementation times, including research and development, are short: six months or less for nearly two-thirds of the source reduction activities for which plants reported implementation time information to INFORM, and six months to three years for another 30 percent.

    - Product yields increased for 97 percent of the source reduction activities in INFORM's study for which information on changes in product yield was provided.

3. Waste-generating facilities are likely to find more source reduction opportunities if they establish source reduction programs with several key features: full **cost accounting** systems, employee training and incentive programs, and high-level leadership that includes both operations and environmental managers. (These program features are discussed in more detail in Chapter 4.) Plants in INFORM's study that adopted any of these program features reported a statistically significant margin in source reduction activities over plants lacking them. Plants that adopted five or more of the eight program features tracked by INFORM reported, on average, two-and-a-half times as many source reduction activities as plants with fewer program features.

4. Government can promote source reduction by enforcing environmental regulations governing the treatment and disposal of wastes. Such regulations have served as an important factor motivating company officials to look for ways to reduce waste: Regulations were the fastest-growing incentive for implementing source reduction activities at INFORM's study plants between 1978 and 1990.

5. Managers of waste-generating facilities can find many low-technology, efficiency-oriented source reduction opportunities when they look for them. Process changes, operation changes, and equipment changes accounted for 87 percent of the source reduction activities reported to INFORM. (These and other source reduction techniques are discussed in more detail in Chapter 3.) It was plants that had previously achieved significant reduction in waste generation that continued to find such efficiency-oriented opportunities.

6. More research is needed to understand the full potential of the more innovative source reduction techniques: product changes and chemical substitutions. A thorough analysis of the benefits, motivating factors, and obstacles associated with these techniques was not possible, as only 13 percent of the source reduction activities in INFORM's study involved such changes. It is worth noting, however, that all of the product changes reported to INFORM with percent reduction data achieved complete elimination of the target waste stream.

Resources

Center for Neighborhood Technology (Valjean McLenighan). *Sustainable Manufacturing: Saving Jobs, Saving the Environment.* 1992. (Center for Neighborhood Technology, 2125 North Avenue, Chicago, IL, 60647. (312) 278-4800. $10 plus $2 postage. Bulk discounts for 10 or more copies.)

Hirschorn, Joel S. and Kirsten U. Oldenburg. *Prosperity Without Pollution, The Prevention Strategy for Industry and Consumers.* New York: Van Nostrand Reinhold, 1991.

Discusses how a preventive environmental strategy makes sense for consumers, workers, government, and business.

The following four books are available from INFORM. Please see page 198 for ordering and discount information.

INFORM. *Cutting Chemical Wastes* (David J. Sarokin, Warren R. Muir, Ph.D., Catherine G. Miller, Ph.D., and Sebastian R. Sperber). 1985. ($47.50)

This study of 29 organic chemical plants in Ohio, New Jersey, and California offers a detailed description of the pollutants generated by and source reduction activities implemented at each facility, as well as a general introduction to source reduction. The case studies will be useful background information as you study plants in your area.

INFORM. *Environmental Dividends: Cutting More Chemical Wastes* (Mark H. Dorfman, Warren R. Muir, PhD., Catherine G. Miller, PhD.). 1992. ($75)

This 1992 study of the same plants covered in Cutting Chemical Wastes *offers a comprehensive analysis of 181 source reduction activities at these organic chemical plants. The analysis includes detailed descriptions of each source reduction activity, quantities of waste reduced, changes in production yield, economic savings, motivation for initiating the source reduction activity, and implementation time. These case studies are an invaluable resource in your efforts to gather and understand information on a specific plant's source reduction methods and progress.*

INFORM. *Tackling Toxics In Everyday Products: A Directory of Organizations* (Nancy Lilienthal, Michèle Ascione, and Adam Flint). 1992. ($19.95)

An information source on toxics in products, including a chart on non-toxic alternatives and an extensive listing of national and international organizations that address toxic product issues.

INFORM. *Preventing Pollution Through Technical Assistance: One State's Experience* (Mark H. Dorfman and John J. Riggio). 1990. ($15)

An evaluation of the North Carolina Pollution Prevention Pays Program, the most well-known state technical assistance program in the country.

The following three publications are available from the National Toxics Campaign at 1168 Commonwealth Avenue, Boston, MA 02134. (617) 232-0327.

The National Toxics Campaign. "The Citizens Toxics Protection Manual." ($35)

The scope of this document is broader than source reduction, but it contains a chapter on toxics use reduction as well as several pointers on obtaining and using information on plants.

National Toxics Campaign. *Fighting Toxics: A Manual For Protecting Your Family, Community, and Workplace.* 1990. ($17)

The scope of this manual is also broader than source reduction, but it contains information on corporate campaigns, neighborhood inspections, working with the media, federal environmental laws, and pollution prevention.

The Good Neighbor Project of the Center for Public Policy. *The Good Neighbor Handbook: A Community-Based Strategy for Sustainable Industry* (Sanford Lewis). 1992. ($25, includes shipping and handling, 15% discount for member groups.)

This handbook addresses multiple issues regarding the promotion of sustainability in industry. It is a resource for developing and implementing "good neighbor agreements."

National Environmental Law Center (William Ryan) and the Center for Policy Alternatives (Rich Schrader). *An Ounce of Toxic Pollution Prevention: Rating States' Toxics Use Reduction Laws.* January 1991. (National Environmental Law Center, 1536 Southeast 11th Avenue, Portland, OR 97214. (503) 231-4181; or the Center for Policy Alternatives, 1875 Connecticut Avenue NW, Suite 710, Washington, DC 20009. (202) 387-6030. $15, plus $2 postage.)

Contains background on toxics use reduction and evaluation of some state toxics use reduction laws, including California, Georgia, Illinois, Indiana, Maine, Massachusetts, Minnesota, Oregon, Tennessee, and Washington. Also contains a recommendation for a model toxics use reduction law.

Northwest Policy Center. *Taming the Toxic Threat: Strategies to Reduce Hazardous Waste Generation in the Northwest* (Kitty Weisman, David Harrison, and Alice Shorett). September 1990. (Northwest Policy Center, 327 Parrington Hall, DC-14, University of Washington Graduate School of Public Affairs, Seattle, WA 98195. (206)543-7900. $10.95 plus $1.50 shipping and handling.)

Office of Technology Assessment. *Serious Reduction of Hazardous Waste.* 1986 (National Technical Information Service, 5285 Port Royal Road, Springfield, VA 22161. (703) 487-4650. NTIS number PB87 139-622. $35 in paperback or $9 on microfiche, plus $3 handling.)

The OTA study of source reduction contains information on everything from definitions to state programs to examples of source reduction to case histories documenting the failure of regulatory programs. Like Cutting Chemical Wastes, *it is a comprehensive introduction to the issue.*

Ohio Environmental Protection Agency, Pollution Prevention Section. *Great Lakes and Provincial Pollution Prevention Programs and the Great Lakes Pollution Prevention Action Plan.* September 1992. (Ohio Environmental Protection Agency, Pollution Prevention Section, Division of Hazardous Waste Management, 1800 Watermark Drive, PO Box 1040, Columbus, OH 43266-1049. (614) 644-3469. $8.25 to cover copying and postage.)

A comprehensive analysis of pollution prevention programs in each of the Great Lakes states and Ontario, including a description of the action plan for pollution prevention in the region.

The Massachusetts Toxics Use Reduction Institute, University of Massachusetts at Lowell. *Toxics Use Reduction Research Directory.* (Toxics Use Reduction Institute, University of Massachusetts at Lowell, One University Avenue, Lowell, MA 01854-2881. (508)-934-3346. Free)

Directory of organizations identified as sponsoring or conducting research on toxics use reduction.

The following EPA publications are available without charge from the Pollution Prevention Information Clearinghouse c/o SAIC 7600-A, Leesburg Pike, Room 369, Falls Church, VA. 22043. (703) 821-4800; or fax your request to (703) 442-0584. Include document number, name, address, and telephone number.)

US Environmental Protection Agency Office of Pollution Prevention and Toxics. *Pollution Prevention 1991, Progress on Reducing Industrial Pollutants.* Document no. EPA 21P-3003. October 1991.

Contains an overview of pollution prevention programs within the EPA and individual states.

US Environmental Protection Agency Office of Pollution Prevention and Toxics and Office of Environmental Engineering and Technology Demonstration. *Pollution Prevention Resources and Training Opportunities in 1992.* Document no. EPA/560/8-92-002. January 1992.

A source of general pollution prevention resources, including documents, manuals, factsheets, reports, videos, state pollution prevention programs, university-affiliated pollution prevention research, training centers, EPA resources such as the 33/50 Program, laboratory pollution prevention, and regional office contacts. The guide also contains educational activities, libraries, the pollution prevention clearinghouse, and an annual calendar of events.

Waste Reduction Institute for Training and Applications Research (WRITAR). *Surveys and Summaries of State Legislation Relating to Pollution Prevention.* (WRITAR, 1313 5th Street SE, Suite 325, Minneapolis, MN 55414-4502. (612) 379-5995. $25 plus $2.50 shipping and handling.)

Contains summaries of pollution prevention legislation enacted in more than 25 states.

Chapter 3 Source Reduction Strategies Plants Can Use

Plant managers wishing to explore source reduction may find opportunities in all areas of their plants. They may find ways to avoid toxic or hazardous chemical use and/or the generation of waste through improvements in the manufacturing process, product changes, better handling of toxic or hazardous chemicals, and more careful chemical storage, loading, transfer, and pollution control operations.

Five Techniques

Source reduction techniques fall into five broad categories: chemical substitutions, product changes, process changes, equipment changes, and operational changes.

1. **Chemical substitutions** involve replacing toxic or hazardous chemicals with nonhazardous or less hazardous ones; that is, using raw materials that create fewer toxic or hazardous by-products in the production process without necessarily changing the product itself, or in conjunction with product changes. A common generic type of chemical substitution is the replacement of toxic organic solvents with nontoxic water-based solvents in a variety of production processes. For example, Ilco Unican Corporation in Rocky Mount, North Carolina, eliminated the use of the toxic organic solvent 1,1,1-trichloroethane for degreasing some of its metal products. The company installed a new system for the degreasing operations that used a water-based biodegradable solution instead of the organic solvent. Savings in 1988 totaled more than $40,000.

2. **Product changes** (product reformulations) involve changes in the product itself that can be achieved without changing the fundamental manufacturing process and that reduce the generation of waste or other potential hazards involved in the manufacture of the product. For example, creating a chemical product in the form of pellets rather than as a powder can reduce the generation of waste dusts as the material is transferred during final packaging op-

erations. This not only reduces waste, but also the potential for worker exposure to the dust. The 3M Company employed a different type of product change: it reformulated a product so that a nonhazardous organic raw material could be used in its manufacture instead of a cadmium alloy, thereby eliminating the need for the metal as a raw material, along with its resultant waste stream.

3. **Process changes** involve any change in the production process that reduces the generation of waste, ranging from simple alterations of process conditions such as temperature and pressure to discovery of new chemical pathways and production technologies. Process changes often improve efficiency (by using fewer raw materials per unit of product) and decrease waste generation and the potential for other toxic hazards that may result from production activities. For example, since 1952 Monsanto's plant in Addyston, Ohio, has gradually changed its polystyrene process from one of batch reactions to a closed-system continuous reaction, to achieve a 99 percent reduction in air emissions. (**Continuous processes** are dedicated to continuous production of a single product, whereas **batch processes** are used to produce a single product at different times or to produce different products at different times to meet changing customer needs. Continuous process operations usually generate less pollution per pound of product than do batch process operations because of fewer start-ups and shut-downs and fewer equipment cleanings.)

4. **Equipment changes** involve modifications of and additions to equipment used in any stage of the manufacturing process (e.g., equipment used for storing, moving, mixing, or reacting chemicals) in order to reduce the amount of waste generated and decrease the potential for other hazards involving toxic chemicals. Because equipment is used in all aspects of a plant's operation, there are numerous opportunities to implement source reduction through equipment changes. An Exxon plant in Bayway, New Jersey, installed floating roofs on 16 tanks storing the most volatile (easily evaporated) chemicals used at the plant. A floating roof is an additional cover that rests on the surface of the liquid stored in the tank. The floating roof rises and falls as the level of the liquid changes, minimizing the formation of chemical vapors that have to be vented from the tank. The Exxon plant's emissions reductions in 1983 totaled 680,000 pounds a year and saved more than $200,000

5. **Operational changes** in the way toxic or hazardous materials are handled at a plant (e.g., careful observation and control of materials, process conditions, and employee habits in order to minimize spills, process upsets, or the use of excessive amounts of chemicals) can reduce the generation of waste and decrease the potential for other toxic hazards. Borden Chemical Company's Fremont, California, plant reduced its phenolic solid waste generation by 93 percent through a series of simple improvements in the handling of phenol. These modifications included saving rinsewaters from tank

and filter washings and converting from a one-stage to a two-stage rinsing of the vats in which chemicals were mixed. By using little water in the first rinse, concentrated chemical wastes could be collected for reuse. Such changes enabled Borden to close down a pollution control evaporation pond at the factory and reduce its annual production of resin sludge from 350 cubic yards to 25 cubic yards, saving $48,750 annually in disposal costs alone.

An important subcategory of operations and/or equipment changes is **in-process recycling** — recycling that is an integral part of the production process. For example, a plant might move a waste stream from one end of an operation to a point near the front end in a closed-loop system for reuse as a raw material in the same operation. Often, a particular process creates a toxic or hazardous substance that can be fed directly back into the process, thus reducing or eliminating the generation of toxic or hazardous waste and increasing the product yield. Aristech Chemicals in Ironton, Ohio, added resin absorption systems to the cumene oxidation process in its phenol production plant. The absorption systems soak up organic vapors and return them to the production process for use as raw materials. This change has prevented 715,000 pounds of cumene vapor releases each year and saved Aristech Chemicals $178,750 annually in cumene costs.

The resources listed at the end of Chapter 2 also are useful for Chapter 3.

When you start to gather information about a plant in your community, you will realize that, as a plant outsider, you probably will never become familiar enough with the details of each production process to identify the most appropriate source reduction opportunities yourself. What you can do is promote the atmosphere and procedures within the company that will enable it to develop an effective source reduction program. If you are in a state that has a source reduction law or voluntary initiative, it may be that such a law or program already requires or encourages a plant to practice source reduction. If so, you can use the following information as a checklist to ensure the effectiveness of the law and perhaps promote a source reduction program at the plant that is even more comprehensive than the source reduction requirements in your state law.

Company Atmosphere

A company that wants to explore source reduction opportunities must broaden its focus from managing to preventing pollution, thereby making its products more efficiently and reducing workplace exposures and environmental emissions. To achieve this goal, companies need to take a variety of steps that often confront ingrained habits and attitudes. INFORM identified these steps in *Environmental Dividends: Cutting More Chemical Wastes:*

- A company must start assessing pollution control costs back to the production processes that use and generate the pollutants, rather than considering them general plant overhead. By assigning these costs back to the processes, companies are better able to grasp the true costs of pollution and seek ways to reduce them. Plants in INFORM's study that did this found an average of three times as many source reduction opportunities as those that did not.

- Companies must examine the continuous opportunities source reduction offers to reduce costs rapidly and increase production efficiency — even for processes that already make commercially successful products. Inertia can set in when the bottom line shows a financially successful product. As noted in Chapter 2, INFORM's *Environmental Dividends* documented that (1) the cost of implementing many source reduction activities is low, (2) companies can often recoup their initial investments rapidly, (3) implementation times, including research and development, are often short, and (4) product yields nearly always increase.

- A company must recognize that everyday plant production workers can play a vital role in identifying source reduction opportunities. The specialty of pollution control departments is pollution treatment and disposal. Their staff members typically have little control over the industrial processes creating the pollution in the first place and little contact with plant managers about the relationship between particular products and the hazards they create. The Office of Technology Assessment (OTA), INFORM, and others have found that ignorance of source reduction opportunities is fed by a culture within many plants (and environmental regulatory agencies) that does not reward the creativity and initiative of the very plant employees who are often best positioned to identify source reduction options: the employees involved in the day-to-day production processes. *(See Box 4-1: The Limitation of Pollution Control Departments.)*

- A company must focus its waste-related data collection on the plant processes themselves, rather than on the end of the pipe, in order to reveal waste sources and opportunities for reduction.

A successful industry program requires changing the company's attitude about source reduction and promoting the flow of information about toxics use, waste generation and pollution problems, costs, and source reduction solutions. It entails developing an atmosphere in which information sharing and creative problem solving are emphasized and rewarded at all levels of the company.

Key Plant Contacts

When you request an interview, be sure you are meeting with those officials who have an understanding of the plant's inner operations, the authority to investigate its environmental performance in depth, and the power to create and implement effective programs.

Now, more than ever, companies are being held accountable by the public for the problems associated with the toxic or hazardous chemicals they use and release as waste into the environment. As a result, many companies have assigned public relations personnel to respond to outside criticism of the company's environmental performance and deal directly with the concerns of citizens and environmental groups. These public relations representatives are usually not as equipped as plant managers and engineers to discuss the plant's inner workings.

While it is crucial to contact plant production managers and engineers, professionals in corporate environmental departments can be important when you start your dialogue. Their job is to ensure the company's environmental compliance with regulations and perhaps to assess opportunities for overall improvement, but they may be knowledgeable mainly about end-of-pipe pollution control measures for reducing waste. However, these types of plant personnel, saddled with the rough job of coping with wastes already created, are becoming more and more eager to promote source reduction at the plant.

Six Steps Companies Can Take

To create an atmosphere in which a source reduction program can flourish, corporate leaders can take six specific steps. They can adopt a source reduction policy; create the proper leadership, including both environmental management and production officials; develop incentives to involve employees; require comprehensive inventories of all sources and activities involving toxic or hazardous materials; collect waste-related data and incorporate it into the cost accounting process; and develop source reduction plans. You will want to know whether the company you study has taken these six specific steps to improve information flow and motivate employees to seek out source reduction opportunities.

1. **Adopt a source reduction policy.** Companies should make source reduction their primary strategy, investigate source reduction opportunities, and make every attempt to reduce the use of toxic or hazardous chemicals and the generation of waste by using one or more of the source reduction techniques discussed in Chapter 3.

    Any company that is serious about source reduction should promote the primacy of source reduction for environmental and workplace protection through explicit written policy. This policy should make clear that source reduction is the preferred option that should be evaluated in all cases, not only for regulated pollutants or particularly costly inefficiencies. Waste management techniques should be considered only after all potential source reduction opportunities are investigated. Such a policy will send a clear message to all employees — not just those with environmental responsibilities — that source reduction is the highest-ranked strategy for addressing problems associated with the use of toxic or hazardous chemicals and the generation of waste.

For example, in 1980, the Du Pont plant in Deepwater, New Jersey, one of INFORM's study plants in *Environmental Dividends*, created a corporate-wide waste reduction policy with source reduction at the top of the hierarchy of waste reduction methods. Beginning in 1983, the plant adopted environmental and employee involvement programs. They include a task force that makes recommendations for quality achievement; committees comprised of environmental coordinators who make waste reduction recommendations and review progress; and monthly meetings on source reduction for all employees. The Deepwater plant reported an annual reduction of more than 39 million pounds of waste, resulting from 10 source reduction activities.

2. **Create combined leadership.** Leadership and oversight of the company source reduction program should be assigned to high-level employees who have responsibility for plant operations and environmental performance. They should have the clout to get things done, and source reduction should be a major factor in their performance evaluations. This step is key to ensuring that the responsibility for running the factory at maximum efficiency with minimum materials loss does not float aimlessly around the company. The directors of source reduction will also play a vital role in improving and coordinating the flow of information about source reduction throughout the company. Of the 29 plants studied in *Environmental Dividends,* INFORM found that those with source reduction leadership from both the environmental department and from the plant manager or other non-environmental departments reported an average of more than twice as many source reduction activities (14.8 activities per plant) as plants with leadership from only the environmental (6.3 activities per plant) or nonenvironmental (6.1 activities per plant) departments.

3. **Develop incentive and involvement programs.** Companies can reward employees who excel at source reduction, just as it rewards those who develop new products or elevate production levels. INFORM's research in *Environmental Dividends* indicates that plants with some type of employee involvement (incentives, training, idea solicitation) had an average of more than twice as many source reduction activities (10.1 activities per plant for partial involvement) as plants with no employee involvement (4.0 activities per plant).

Incentives such as awards or bonuses can be powerful tools for combatting natural employee tendencies not to tamper with the status quo. Companies can also hold contests whereby employees submit source reduction proposals, and winners are announced at a public ceremony. For example, the "Waste Reduction Always Pays" (WRAP) program created by Dow Chemical is a recognition and reward system for employees who suggest source reduction ideas that reduce waste and costs. Dow reported that the competitive nature of a recognition and reward system has encouraged employees to look for source reduction opportunities. Rewards and recognition include articles in company newspapers, plaques, recognition dinners, and trips to headquarters or other locations to meet with senior management. In 1986, Dow held a contest in its Louisiana Division as part of the WRAP program. Twenty-three projects costing $3.3 million to implement were selected: within a year, they saved Dow $6.4 million, for a 193 percent return on investment.

4. **Conduct a full inventory of all sources and activities that involve toxic or hazardous substances.** By conducting a complete audit of every source or activity within the plant that involves toxic or hazardous substances, companies can get a full picture of where toxics are used and lost within their facility. Research by INFORM and others, including OTA, has found widespread potential for reducing hazards from all stages of plant operations: during handling, storage, loading, transfer, and pollution control operations, as well as from production processes and products.

5. **Incorporate waste-related data into the plant's cost accounting system.** Before plant managers can develop effective source reduction plans for their facilities, they need to both track the sources of their waste and identify the costs associated with each waste source, thereby making it possible to assign the waste-related costs back to the process that creates them. They need three types of data to identify and prioritize source reduction opportunities and track source reduction achievements: facility-wide waste information, process-specific waste information, and cost data.

First, managers need data on the movement and fate of chemicals used plantwide in order to measure the plant's overall source reduction progress. They need to know how much of each chemical enters the plant; how much is consumed in production; how much leaves the facility for disposal, recycling, or in the product; and how much is released as waste.

This plant-wide information is obtained through a facility-level **materials accounting** (tracking of raw materials and products from one end of the plant to the other). For each chemical raw material, managers compare the sum of the amount of the material already existing in plant inventory and the material entering or produced in the plant, with the sum of the amount of material consumed in plant processes and the amount shipped out of the plant in products. The difference between these sums, within measurement accuracy, is the amount of the chemical that becomes waste. (For an explanation of measurement accuracy, see the discussion of Worksheet 10, Section D in Chapter 5.)

Second, a plant manager needs information about the movement and fate of chemicals within each process in order to identify specific sources of waste and activities leading to waste generation. This allows plant and process managers to consider what changes can be made to alter or eliminate these sources. Such information is obtained through an inventory that includes, on a process-by-process basis, the same kind of comparison of material inputs and outputs that is done on a facility-wide basis, and a detailed analysis of how materials are handled and wastes generated within each process.

Developing this process-level information base requires a high-level employee with a background in production operations, not pollution control, to direct annual source reduction inventories and develop plans for each production line in the company. As such a data collection process has many steps, and changes recommended as a result of the inventory could affect other departments within the company, the project director would probably want to create a study team that includes employees involved in many aspects of the company's work. The team should include representatives from facility process and environmental engineering, safety and health, purchasing, materials and inventory control, maintenance, finance, marketing, and product quality control. The team:

- Conducts a comprehensive inventory of all toxic or hazardous materials used or produced in the plant.

- Collects existing data on facility layout, waste stream generation, and waste-related cost accounting, and background data on each production process, e.g., schematic diagrams.

- Conducts a plant survey to verify background data and identify major additional waste streams.

- Samples waste streams to determine quantity and composition if this information is unavailable.

- Conducts a quantitative assessment, a materials accounting, for each process to make sure that all major sources of waste generation (regardless of environmental medium) are measured and accounted for.

- Collects data on actual operation and management practices. This **process-level source reduction inventory** is a series of steps enabling internal company management to identify the sources and activities, on an individual process basis, that lead to the use of toxic or hazardous substances or the generation of waste. This inventory must not rely solely on the quantitative materials accounting of input and output. Relatively small levels of waste can be identified only through actual measurement at possible sites of release and through asking the operations personnel questions designed to determine the operator activities

Box 4-2: Cost Accounting

Usually Includes:
- Labor
- Fuel
- Raw material inputs
- Capital
- Operating

Should Also Include:
- Regulatory compliance
- Lost materials
- Liability
- Pollution control
  and disposal

and status of equipment that might lead to the generation of waste materials in small quantities.

Third, the manager should ensure that waste-related costs are charged directly to the processes that create the waste, rather than to general plant overhead, thus demonstrating that waste adds to the costs of production. This involves assigning to each process the value of any lost raw materials and products revealed by the source reduction inventories, along with the cost of managing any waste created by that specific process.

Companies use cost accounting principles to measure the raw materials, capital, operating, and labor costs of a specific production line. *(See Box 4-2: Cost Accounting.)* These principles allow companies to balance the cost of production against sales of a given product and maximize profit. If waste costs are factored into this cost accounting, the plant manager has a more accurate estimate of the true cost of manufacturing the product. This procedure helps bring the environmental engineers "inside" the plant, and brings home the importance of reducing toxics use and preventing waste generation to production managers and their employees. Like rewards, this helps ensure that each employee's "bottom line" reflects the overarching priority of reducing waste generation at the source.

6. **Develop plans for source reduction activities in the plant.** Upon completion of the inventory and cost accounting, a manager should be able to list the production processes and waste streams in priority order for source reduction and to identify applicable techniques. More information about the techniques described briefly in Chapter 3 is available from state technical assistance programs, from the EPA Pollution Prevention Information Clearinghouse, and from private consultants and chemical suppliers.

   Individual strategies for source reduction may best be developed by a company source reduction team's own brainstorming, as source reduction opportunities are often very plant-specific. Thus, it is essential to share the results of each inventory with all the operators of the production line in question. In many cases, they will not have had this kind of information before, but they may be best-equipped to use it effectively and may have additional suggestions for source reduction initiatives.

Ideas for making production more efficient should be selected based on their potential to prevent environmental, occupational, and public health damage, their feasibility, and their costs. The selected projects should be organized into source reduction plans, forming a working agenda for source reduction in facilities and company-wide. They should contain goals and timetables for source reduction achievements, and their implementation should be closely monitored.

Resources

Dow Chemical Company. (Department of Environmental Quality, Building 2030, Dow Chemical Company, Midland, MI 48674. (517) 636-6999. Free.)
Dow has written several short descriptions of its source reduction program.

The two following publications are available from INFORM. See p. 198 for ordering and discount information.

INFORM. *Cutting Chemical Wastes* (David J. Sarokin, Warren R. Muir, Ph.D., Catherine G. Miller, Ph.D., and Sebastian R. Sperber). 1985. ($47)
This study of 29 organic chemical plants in Ohio, New Jersey, and California offers a detailed description of the pollutants generated by and source reduction activities at each facility, as well as a general introduction to source reduction. The case studies will give you useful background as you study plants in your area.

INFORM. *Environmental Dividends: Cutting More Chemical Wastes* (Mark H. Dorfman, Warren R. Muir, Ph.D., Catherine G. Miller, Ph.D.). 1992. ($75)
This 1992 study of the plants covered in Cutting Chemical Wastes *offers a comprehensive analysis of 181 source reduction activities at these organic chemical plants. The analysis includes detailed descriptions of each source reduction activity, quantities of waste reduced, changes in production yield, economic savings, motivation for initiating source reduction activity, and implementation time. These case studies are an invaluable resource in your efforts to gather and understand information on a specific plant's source reduction methods and progress.*

The Massachusetts Toxics Use Reduction Institute, University of Massachusetts at Lowell. *Toxics Use Reduction Research Directory.* (Toxics Use Reduction Institute, University of Massachusetts at Lowell, One University Avenue, Lowell, MA 01854-2881. (508)-934-3346.)
Directory of organizations identified as sponsoring or conducting research on toxics use reduction.

The following two books are available from the National Toxics Campaign, 1168 Commonwealth Avenue, Boston, MA 02134. (617) 232-0327.

National Toxics Campaign. "The Citizens Toxics Protection Manual." ($35)

> *In addition to contents described in Resources list for Chapter 2, this book contains a useful discussion of important elements of an industry source reduction plan in Chapter 10.*

National Toxics Campaign. *Fighting Toxics, A Manual for Protecting Your Family, Community, and Workplace.* 1990. ($17)

> *In addition to contents described in Resources list for Chapter 2, this book contains a useful discussion of important elements of an industry source reduction plan in Chapter 9.*

Office of Technology Assessment. *Serious Reduction of Hazardous Waste.* (Summary available from the Office of Technology Assessment, United States Congress, Washington, DC 20510. (202) 224-8996. Free. Book available from the National Technical Information Service, 5285 Port Royal Road, Springfield, VA 22161. $28.95 in paperback or $6.95 on microfiche, plus $3 per order for handling for either edition.)

> *The OTA report contains an extensive discussion of corporate source reduction.*

US Environmental Protection Agency. *Pollution Prevention Information Clearinghouse.* (Call (800) 424-9346.)

> *The Clearinghouse can provide information on source reduction techniques.*

US Environmental Protection Agency, Risk Reduction Engineering Research Laboratory. *Facility Pollution Prevention Planning Guide.* 1992. Document EPA/600/R-92/088. (Center for Environmental Research Information. (513) 569-7562. Free.)

> *This guide summarizes the benefits of a company-wide pollution prevention program and suggests ways to incorporate pollution prevention in company policies and practices.*

Chapter 5   Surveying Your Local Plant

This chapter will guide you step-by-step through the process of studying individual plants to determine whether they are using toxic chemicals and releasing them as waste, to assess whether they have made a commitment to source reduction, and to evaluate the results. This process consists of four stages :

1. Conducting research on the plant.

2. Preparing for the plant interview.

3. Interviewing plant representatives.

4. Conclusions and follow-up.

The worksheets at the end of this chapter list the questions you will be asking in both the plant research and plant interview phases.

Conducting Research on the Plant

Considering Which Plants to Study

You may already know which plant or plants in your community you would like to assess. If not, you may consult a variety of sources to obtain information to help you make a selection. *(See guidance for Worksheets 1-3.)* In most cases this information will help you identify a plant that is significant to your community, whether because of its size, the amount of environmental pollution it creates, or some other factor.

Conducting Plant Research

The following suggestions may be helpful as you conduct research on the plant(s) in your community.

Read this chapter completely before you begin the plant research. This will help you gain a better perspective on the kinds of information available on most plants and help you assign priorities to the information, based upon your interest and resources. For example, if you suspect the plant has many toxic emissions or a high chemical accident rate, you may wish to concentrate more of your efforts in these more troublesome areas.

Do the research thoroughly and carefully before you contact plant officials. Your discussions with company representatives will be most effective if you have learned much about the plant and its activities before you meet. Your preparation prior to the plant interview is important for two reasons. First, conducting background research on the plant and understanding the information collected provides some insight into the plant's toxic or hazardous chemical problems. Second, you can more easily assess statements by company representatives about the quality of the plant's environmental or safety records if you have this information prior to the plant interview.

Gather as much data on the plant as you can through public data sources, using Worksheets 1-9. Always keep your information organized and save the names of contacts and sources of data. As you conduct your research, you should be open to other questions and issues that may arise as your inquiries are answered: an answer to one question may lead to another question, and so on. Always ask for clarification of an answer that you do not understand. You may also ask public officials where to find other information on the subject area, or whether there are others in the field who would be useful to contact. You may enlist the help of neighbors or members of environmental, labor, or public interest organizations in your community to assist you with this information search. Finally, you may consult information hotlines and publications listed in the Resources at the end of this chapter.

Obtaining and understanding plant background information is hard work. Indepth research on a plant can be time-consuming and frustrating: It can take weeks, or even months, depending on federal and state agency responses and your resources, to obtain some or all of the information you seek. As you conduct your search, you may encounter uncooperative government officials or wait several weeks to get answers. You may have to make several inquiries before you get the information you want. You may find that some of the information does not exist or is out of date for the plant in your community. In some cases you will have to pay a small fee for information or the cost of copies. This guide focuses on information from data sources that are especially useful to your efforts to promote source reduction; there are dozens of other environmental and toxics-related databases you may consult.

Finally, conducting a background check on a company and engaging in dialogue with plant operators will not solve all of the environmental problems in your community, nor even all of the problems at a particular plant. It will, however, provide you — a neighbor and citizen who has the right to know about environmental and health hazards where you live — with important information that may enable you to make a strong, positive impact on health and environmental quality in your community.

Plant Research Guidance

The following sections separate publicly available information into specific areas. Each section has a corresponding worksheet to help you organize the information and formulate useful questions about the data you are collecting. The availability of this data may vary depending upon the industry, plant size, number of employees, or other factors. If you are unable to find the information through public sources, you can question plant officials about these issues after you have conducted the source reduction interview discussed later in this chapter. (Worksheets 1-11 are on pages 89 through 123.)

Please note that in some cases you may need to photocopy the worksheets in order to fill in information on multiple chemicals or incidents.

Worksheet 1 — Plant Identification and Profile

The first step is to gather general information about the characteristics of the plant you are studying. This information is important for four reasons. First, data about production, employment, sales volume, and products can help put the company's program in perspective. It may be much more impressive for a small company to eliminate 100 tons of a given contaminant than for a larger one to achieve the same result. Second, obtaining such information could prove invaluable in comparing the performance of one plant with others that share the same characteristics. Third, it is important to determine who actually owns the plant in order to identify who is responsible for the company's environmental policy. Finally, your research on the product(s) manufactured at the plant will provide a greater understanding of why the plant operates in a certain manner or has a particular type of production process. For example, continuous production processes usually generate less waste per pound of product (because equipment does not need to be cleaned as often).

The following sources can help you complete Worksheet 1:

- Company annual reports (available from the company).

- State manufacturing directories. The *Standard Industrial Classification Manual* contains **Standard Industrial Classification (SIC)** codes: the system the Federal government uses to classify US companies according to the products they produce. The *Thomas Register* is a supplement to the SIC manual that contains a company index and information about specific products manufactured by a company. Further, it lists the company's competitors and often contains pictures of the company's product(s). Both publications are available from the US Department of Commerce or the Chamber of Commerce in your state, from public libraries, and from the National Technical Information Service. (For more information on SIC codes and the National Technical Information Service, see Table 5-1 and the section on the Toxics Release Inventory in the discussion of Worksheet 3.)

- Dun & Bradstreet reports. These financial reports on individual companies are available to subscribers; you may obtain information from the Dun & Bradstreet office in the nearest large city, or call (800) 362-2255 to become a subscriber. You will be charged a fee for these reports. The facility's Dun & Bradstreet number will be provided to you over the phone. Remember that there is a different number for the parent company and each subsidiary facility.

- Securities and Exchange Commission (SEC) 10K reports. These reports contain general information on a company's operations, including ownership, sales data, and date operations began. They are available from local SEC offices or SEC headquarters in Washington: 450 5th Street NW, Washington DC 20549. You may be charged a fee for the copies. 10K reports may also be available from an investor relations or similar service department of the company itself and most public libraries.

Worksheet *2* Plant Pollution Profile: Permit Compliance and Applications

Information about the extent to which the plant has met the requirements of the various permits controlling its management or release of toxic or hazardous waste to the air, land, or water can be useful in several ways. First, if you find evidence of a major violation or pattern of violations, you can use it to help convince the state and local community to persuade the plant to adopt a comprehensive source reduction program. Second, you may be able to convince the plant manager that compliance problems can best be avoided through a program that reduces the quantity of chemical releases that must be managed under complex permit conditions. Third, circumstances are often most favorable for persuading a plant to adopt a source reduction plan when it is seeking approval from the state for a new permit or for some change in its existing permit conditions. Most states have procedures that provide for a public hearing, or at least an opportunity for public comment, on a company's application for a new or modified permit.

Remember, however, that source reduction extends to all chemical releases, regardless of whether they are subject to permit requirements. For example, many chemicals included in the Toxics Release Inventory are regulated only when discharged into a particular environmental medium, or not at all.

Worksheet 2, including the table on specific violations, can be completed by reviewing permit files at the state environmental office or by speaking directly to the officials in that office who write the permits. These officials, who actually wrote the pollutant discharge permits for the plant you are studying, are some of your best sources of information. State agencies are usually divided into separate departments for permits for different environmental media, such as an air permitting office, a division of water permitting, etc. The permit writers in these offices can help you understand how the permitting system operates, or provide more information about the specific laws or regulations that give them permitting authority. Most important, they can provide permit information about specific facilities.

As there may be a different permit writer for each environmental medium (i.e., air, water, and land), you may need to speak with more than one person. If you need help in identifying the appropriate contact in your state agency for information regarding the permit(s) in the plant you are researching, contact the following professional organizations:

Solid Waste (Land Disposal)

Association of State and Territorial
Solid Waste Management Officials
444 N. Capitol Street, NW, Suite 388
Washington, DC 20001
Tel. (202) 624-5828
Attn: Nicole Malloy

Water Discharges

Association of State and Interstate
Water Pollution Control Administrators
444 N. Capitol Street, NW, Suite 330
Washington, DC 20001
Tel. (202) 624-7782

Air Emissions

Waste Management Association
PO Box 2861
Pittsburgh, PA 15230
Tel. (412) 232-3444

State Air Pollution Control Administrators
444 N. Capitol Street, NW, Suite 306
Washington, DC 20001
Tel. (202) 624-7864
Attn: Nancy Kruger

Once you reach the appropriate state contact, you may want to make an appointment to visit the office and review the plant's permits. In any event, if the state office is not nearby, you can obtain information over the phone or request written materials.

The availability of information on state permits and permit compliance varies from state to state. In some cases, agency officials may be willing to speak with you about information on a specific plant. In other cases, they may ask you to submit a request in writing or visit the state office and review the materials on your own. Much of the information you gather on permit compliance and applications will depend upon your ability to ask questions and probe state officials for more information. If the official asks that you make a Freedom of Information Act

(FOIA) request, see Appendix F for pointers on how to do so. Use Worksheet 2 as a guide in your information request.

If you plan to be involved in surveys in more than one state, you may wish to purchase the membership directories that each of the above organizations produces. However, in some cases these directories list only the administrator or main office for the program and may not provide the name and address of the individual responsible for information about specific facility permits.

Worksheet 3 — Plant Pollution Profile: Toxics Release Inventory

In 1986, the United States Congress passed the Emergency Planning and Community Right-to-Know Act (Title III of the Superfund Amendments and Reauthorization Act.) This important law provides citizens with the right to find out about the toxic or hazardous chemicals stored, used, and released in their communities. In particular, Section 313 of Title III created the Toxics Release Inventory (TRI) to provide public data on "routine" chemical releases from industries across the United States. With certain limitations, the Toxics Release Inventory can provide you with specific information on how much of which chemicals each plant in your community is releasing to each segment of the environment. Use section A of Worksheet 3 to collect this information.

In October 1990, Congress passed the Pollution Prevention Act. This law requires a new emphasis on pollution prevention and source reduction among US Environmental Protection Agency officials and covered industries. It requires all facilities that currently submit TRI information to also include specific information on energy recovery, recycling, treatment, and source reduction. This newly available public information has recently been added to the same reporting package as the TRI information. Section B of the Worksheet focuses on these data.

In 1991, EPA initiated the 33/50 Program, which asked companies to voluntarily reduce releases and off-site transfers of 17 specific TRI chemicals by 33 percent by the end of 1992 and 50% by 1995. Participating companies are encouraged, but not required, to use source reduction strategies to meet these goals. If the plant you are studying has joined the 33/50 Program, you will be able to find out what it has committed itself to do. Section C of Worksheet 3, discussed later in this chapter, will guide you through collecting the available information.

About the Toxics Release Inventory

Understanding what the TRI is and what information it contains is vital to using the data effectively in working with local companies and state government to maximize source reduction. Similarly, it is useful to understand how this information is enhanced by the new information requirements in the Pollution Prevention Act of 1990. This section gives an overview of the publicly available data.

Appendices A and B provide the state contacts for gathering TRI and pollution prevention information. Appendix C lists the TRI chemicals and their health effects.

How Does the Toxics Release Inventory Work?

Before you begin to collect the information in Worksheet 3, it is important that you understand the nuts and bolts of the TRI.

The TRI legislation requires that all companies meeting certain criteria (described in the next section) report annually to the US Environmental Protection Agency how much of each of certain specified chemicals they released to each environmental medium: air, water, and land. The chemicals that must be reported include those that are acutely lethal, possibly carcinogenic, capable of significant adverse effects, or mildly toxic. (The original TRI chemical list specified 308 individual chemicals and 20 chemical categories. However, this number occasionally changes as chemicals are added to and deleted from the list; as of early 1993, there were 302 individual chemicals and 20 chemical categories.)

EPA is required to make this information available to the public. The first inventory, containing 1987 data, was released in mid-1989; it was the first time that specific figures on individual chemical releases from individual industrial plants to individual environmental media were gathered and publicly disseminated. As of early 1993, the most recent TRI data available contained 1991 figures.

Which Companies Must Report?

A company must provide data on its chemical releases to EPA if it: has 10 or more full time employees; is a manufacturing industry listed within Standard Industrial Classification (SIC) codes 20 through 30 *(see Table 5-1);* and manufactures, imports, or processes more than 25,000 pounds of the listed chemicals, or uses 10,000 or more pounds per year of one or more of the listed chemicals. Appendix C lists the chemicals and chemical categories that companies must report, both alphabetically (Table C-1) and by CAS number (Table C-2). The American Chemical Society assigns the **CAS** or **Chemical Abstract Services** number to each chemical to identify accurately chemicals that may have several names. The SIC Manual may be found in most libraries, or a copy may be ordered from:

> The National Technical Information Service
> 5285 Port Royal Road
> Springfield, VA 22161
> Tel. (703) 487-4650

If you find the SIC Manual in a library, you may also want to look for the *Thomas Register* (described in the discussion of Worksheet 1).

Table 5-1: Industries Required to Report TRI Data
(by Standard Industrial Classification Code*)

SIC Codes	Industry Group
2011-2099	Food and kindred products
2111-2141	Tobacco manufacturers
2211-2299	Textile mill products
2311-2399	Apparel and other finished products made from fabrics and other similar materials
2411-2499	Lumber and wood products (except furniture)
2511-2599	Furniture and fixtures
2611-2661	Paper and allied products
2711-2795	Printing, publishing, and allied industries
2811-2899	Chemicals and allied products
2911-2999	Petroleum refining and related industries (e.g., coal)
3011-3079	Rubber and plastic products
3111-3199	Leather and leather products
3211-3299	Stone, clay, glass, and concrete products
3311-3399	Primary metal industries
3411-3499	Fabricated metal products (except machinery and transportation equipment)
3511-3599	Machinery (except electrical)
3611-3699	Electrical and electronic machinery, equipment, and supplies
3711-3799	Transportation equipment
3811-3873	Measuring, analyzing, and controlling instruments; photographic, medical, and optical goods; watches and clocks
3911-3999	Miscellaneous manufacturing industries

* The Standard Industrial Classification Manual categorizes the products produced by industrial facilities. Industries are classified by two-, three-, and four- digit SIC codes. The two-digit code identifies the major group as a whole, the three-digit code provides a more specific sector within that larger group, and the four-digit code pinpoints a very specific industry. The four-digit code will be the most useful for identifying the type of product or service provided by the plant you are investigating. For example, the two-digit code 28 indicates the chemical and allied products industry, the three-digit code 281 indicates that the chemicals are industrial inorganic chemicals, and the four-digit code 2812 indicates that the specific inorganic chemicals in this category include alkalies and chlorine.

What Information Is Reported?

Companies reporting Toxics Release Inventory data fill out EPA Form R, the Toxic Chemical Release Reporting Form, for each chemical (a blank form is included in Appendix D). The form originally asked only for information on the company and its chemical releases. With the passage of the Pollution Prevention Act, the same companies must now also report recycling and source reduction data on Form R.

The Toxics Release Inventory Sections of Worksheet 3

Section A: Total Releases to the Environment On-Site and Off-Site

Before you meet with company officials, you should know what hazardous and toxic materials the plant is releasing and transferring off-site, and where these releases and transfers are going. Section A of Worksheet 3 is designed to help you organize and understand this information. It provides space to list the amounts of each pollutant released to the different segments of the environment. As long as the company you are investigating is a manufacturer (not an incinerator, gas station, or some other nonmanufacturing operation) and meets TRI reporting requirements (such as size and amounts of chemicals used), the TRI Form R is your best source of information on toxic releases to the environment. This information, supplemented by the state permit compliance information (Worksheet 2) and the supplemental right-to-know data on chemicals and chemical storage (Worksheet 4), is a key component of your research on a plant.

Section B: Recycling and Source Reduction

Besides learning about plant releases, you may also use Toxics Release Inventory information to investigate the plant's facility-wide waste management and source reduction performance. The data include quantities of hazardous materials used for on- and off-site **energy recovery** (obtaining energy from the incineration of solid waste); recycled on- and off-site; treated on- and off-site (**hazardous waste treatment** entails changing the composition of the waste to render it non-hazardous); and released into the environment as a result of remedial actions or catastrophic events. Form R also includes a **production ratio or activity index** (a method for measuring the quantity of waste generated in relation to the quantity of product) and information about all source reduction activities undertaken during the reporting year.

As a source reduction program should be designed specifically to reduce the quantity of materials entering the waste stream, you need to know the amount of all toxic or hazardous wastes actually generated during production, before any kind of out-of-process recycling, treatment, or disposal. Remember that source reduction means avoiding the creation of a waste in the first place. This information can be obtained through simple calculation, by adding numbers 8.1 through 8.8 in Section 8 of Form R. (As of early 1993, it was unclear whether EPA would be making this calculation for you.) EPA has produced a useful booklet, *A Guide to the Toxic Chemical Release Inventory Form R,* which contains a section-by-section description of data collected on the form (see Resources at the end of this chapter).

How To Obtain Toxics Release Inventory and Pollution Prevention Act Source Reduction Information

TRI and Pollution Prevention Act source reduction information is available in three ways: by obtaining copies of the Form Rs the plant has submitted (one for each reportable substance each year); by obtaining computer printouts or computer disks containing Form R data; or by using a computer and a modem to obtain the information through publicly available databases.

Toxics Release Inventory Access without a Computer

If you do not have a computer and wish to obtain printouts of the data contained in the TRI Form Rs submitted by the plant you are studying, you can start by contacting your state TRI representative (listed in Appendix A). The Information Management Office at the US Environmental Protection Agency also has these records. As the Form Rs are submitted every year, you should ask for the information for the most recent year (this will also ensure that you receive the new source reduction data). Geographical reports (containing TRI data by zip code) are available, allowing you to assess the TRI emissions in a larger area. You may also request full copies of the Form Rs, but remember that each Form R is several pages long, and most companies submit several forms to the agency. When you request the data, be sure to ask the TRI representative to notify you of the size of the request and the cost, if any, before your order is filled.

EPA Office of Pollution Prevention and Toxics
Information Management Division
401 M Street, SW
Washington, DC 20460
Tel. (202) 260-6238

If you have further questions on the TRI, more information is available from EPA's Emergency Planning and Community Right-to-Know Act (EPCRA) Information Hotline at (800) 535-0202 [or (703) 412-9877 in Virginia or Alaska]. TRI data are also available in a public reading room at EPA headquarters.

The state contacts (listed in Appendix A) can tell you how to obtain the information at a state location. Finally, the company you are interested in may be willing to give you information, although no company is required by law to distribute its TRI forms. Many libraries are beginning to carry copies of the TRI data.

Computer Access to the TRI Data

EPA is required by law to make the TRI data publicly accessible via telecommunications. It has contracted to do so through the National Library of Medicine's TOXNET system. The system is available 24 hours a day, seven days a week, to anyone with a computer, communications software, a phone line, and a modem. Call (800) 638-8480 for information on how to subscribe to TOXNET. The on-line fee is $18.00 per hour. If you wish to obtain off-line printouts of the TRI information, you may request them at 30 cents a page.

Information in the database is categorized by company, chemical, medium to which the chemical is released, waste treatment, and off-site transfer. Thus, the system can enable you to answer such questions as: How much of chemical X did company Y release at facilities throughout the nation last year? What are the names and addresses of steel plants importing chemical X in city Y? What quantity of chemical X did company Y ship off-site last year?

Information about TOXNET is available from:

> TRI Representative
> Specialized Information Services
> National Library of Medicine
> 8600 Rockville Pike
> Bethesda, MD 20894
> Tel. (301) 496-6531
> User Support (800) 231-3766

Access to the TRI data is also available through RTKNET, the Right-to-Know Computer Network based in Washington, DC, and run jointly by OMB Watch and the Unison Institute (two organizations that specialize in TRI data). There is no cost for using RTKNET other than the phone call, although organizations in financial need may apply to use a toll-free number. RTKNET also provides communications software to access this database. RTKNET may be willing to do a one-time search free of charge and send you a print-out of the data. However, for any regular use, they recommend that you access the network yourself.

Five types of information are available on RTKNET:

1. Toxics Release Inventory data from 1987-90 on US companies discharging toxic or hazardous chemicals to the air, water, land, and off-site facilities.

2. Fact sheets prepared by the New Jersey Department of Health on 320 toxic chemicals. Many of these sheets are available in Spanish. (See Appendix C for information on how to contact the NJ Department of Health directly.)

3. The US Environmental Protection Agency FINDS database, or Facility Index Data System, which provides a way to find facilities that are regulated under a variety of EPA programs by assigning a unique ID number to each facility. If RTKNET indicates that additional information on the facility (other than TRI data) is available in another database, you can use the unique ID number to request the records; in most cases you will have to file a Freedom of Information Act request. (See Appendix F for the Freedom of Information Act EPA regional contacts.)

4. Company or issue profiles: RTKNET will locate all its available information about a company or a chemical, and you can compile this on-line into a profile.

5. The DOCKET database, which includes details on all the civil lawsuits brought by EPA since its inception. You can find out about environmental infractions of a specific company, facility, or location, or about infractions under a specific piece of legislation, such as the Clean Air Act.

Contact RTKNET at:
1731 Connecticut Avenue, NW
Washington, DC 20009-1146
Tel. (202) 234-8494
Fax (202) 234-8584

How To Use Toxics Release Inventory and Pollution Prevention Act Source Reduction Information

The TRI (and source reduction) database gives you access to information about many chemicals at many plants. Thus, it enables you not only to study an individual plant and its releases, but also to compare plants and use this comparison to decide on which plants and chemicals you will focus. It will also provide you with a ready reference when meeting with industrial representatives.

Once you have the data organized, you can use them to identify target chemicals for reduction at individual companies. You may use the following criteria for targeting chemicals:

Amount of individual chemicals released to the environment annually. You may compile data on releases of individual chemicals from all companies in your area. If several companies are releasing the same chemical, there may be a cumulative health and environmental risk that none of the companies has considered.

Degree of threat to public health and environment posed by individual chemicals. Appendix C gives an overview of the known health effects of each of the TRI chemicals. Upon request, the EPA Right-to-Know Hotline will provide hazardous substance fact sheets for individual TRI chemicals (see the Resources section at the end of this chapter). As the chemicals on the TRI list vary widely in the degree of hazard they pose, it is useful to sort out potential major problems from minor ones.

Comparison with other companies that produce the same product. Using TOXNET or RTKNET, you can compare the emissions of any company to those of all other US companies in the same industrial category. If a plant in your community is generating a lot more waste than other companies of its type, you can bring this to the attention of the company representatives you meet and ask why others seem to be operating more efficiently. Although there may be differences in waste generation due to the size of the plant, and unavoidable differences in efficiency due to genuine product differences, it is also possible that other companies have identified more efficient methods of operation. In the latter case, you will be doing the company a favor by bringing the discrepancy to its attention. In either case, as someone who must live with the pollution, you deserve an explanation.

Identification of rivers and streams in the area receiving chemical discharges. By studying TRI data for all industries in the area, you may find that certain bodies of surface water are receiving large amounts of TRI wastes. You can bring this information to the attention of all companies involved, or to the state environmental protection agency, and seek reductions.

Comparison of current year's chemical emissions with the previous year's emissions. As of early 1993, TRI data are available for 1987-1991 reporting years. You will be able to figure out whether emissions have increased or decreased from year to year for any chemical the company reports under TRI. Likewise, you may want to ask company representatives for graphs of overall emission trends for land, water, and air pollution, or for specific chemicals of concern. This information enables you to study the patterns of pollution at the plant over a number of years. Be careful, however. Increases and decreases can be a result of any number of factors. Overall increases in waste are clearly a concern and should be explained by the company. Did the levels of production increase? Has a new product line been added? Although overall decreases in waste generation may be a positive sign, they may not indicate source reduction. Increased on-site treatment and recycling can decrease emissions levels just as source reduction does. However, getting an overview of pollution trends at the company will be helpful even though it can lead to further questions. If you ask for an explanation of how the reductions were achieved, you can distinguish between true waste reduction and "paper changes" or "phantom reductions."

Indication of source reduction activities. Adding the figures on Form R in Section 8 (numbers 8.1 through 8.7) will provide the total quantity of the hazardous substance generated prior to recycling, treatment, energy recovery, or disposal. The sum will give you the figure that should be targeted for reduction. This figure represents the true quantity of facility-wide waste generation of that substance. If companies are working toward reduction of that number, you will know that their source reduction efforts are at least headed in the right direction.

What Information the Toxics Release Inventory and Pollution Prevention Act Do Not Provide

It is important to understand that the Toxics Release Inventory does not provide information about all wastes generated and released into the environment, even for facilities that report to the TRI. Only some 300 chemicals are targeted, and companies need only report releases and transfers of these targeted chemicals if their manufacture, production, or use exceeds threshold levels. Small releases may still be of concern if they create high exposure levels, such as through direct ingestion of drinking water containing chemicals released by several different companies. Further, only facilities in the manufacturing sector must report. Thousands of other industrial and non-industrial waste-generating facilities (including federal facilities*) are not required to report.

In 1991, INFORM published an analysis of how the TRI could be expanded to provide the public with more complete information about toxic chemicals released to the environment: *Toward a More Informed Public: Recommendations*

---

* *In April 1993 President Clinton announced that he would sign an executive order requiring Federal facilities to report to the Toxics Release Inventory. The EPA said it would consider expanding TRI to include additional chemicals and industrial sources.*

for Improving the Toxics Release Inventory. The following list highlights a few of the most important ways in which the Toxics Release Inventory could be expanded.

- **Include more chemicals.** Many potentially dangerous chemicals are not on the TRI list: for example, various chemicals listed as known or probable carcinogens by the US Department of Health and Human Services and the International Agency for Research on Cancer (IARC) are not covered by TRI reporting requirements. Thus, TRI does not necessarily alert you to all the chemicals of concern released by a particular company. Although there have been some new chemicals added to the list, it is still relatively small. In late 1992, EPA announced its intention to exercise its authority under the law to expand the number of chemicals covered.

- **Include more types of facilities.** TRI covers only plants in the general manufacturing industries, such as chemical manufacturers and petroleum refineries. Nonmanufacturing facilities such as dry cleaners, power plants, and gas stations generate sizeable releases of toxic or hazardous chemicals. Moreover, **waste treatment, storage and disposal facilities (TSDs)** are not required to report.

- **Include information on the movement of chemicals through a plant.** The TRI reporting requirements do not include materials tracking data. Doing so would enable the company, government agencies, and the public to assess quantities of toxic or hazardous substances going in and coming out of the plant. (See the discussion of materials balance in the guidance for Worksheet 10, Section D.)

- **Include process-specific information and toxic or hazardous chemical use information.** The TRI currently requires facilities to report on a facility-wide basis, not on a process-specific basis, and to report releases, rather than uses, of toxic chemicals. Several states, such as Massachusetts and New Jersey, have enacted laws that require companies to audit their chemical use within each production process and develop plans for reductions.

Just because information is not publicly available does not mean you cannot ask for it; you may want to ask plant representatives to provide some or all of the information listed above. For example, you can ask for data on chemicals that are not covered by TRI or chemicals released in quantities below the TRI threshold.

The US EPA 33/50 Program Section of Worksheet 3

The 33/50 Program is a US Environmental Protection Agency pollution prevention initiative that asks companies to voluntarily reduce releases and off-site transfers of 17 specific toxic chemicals. The goal of the 33/50 Program is to achieve a national average of 33 percent reduction in the release of these chemicals by the end of 1992 and 50 percent reduction by the end of 1995. EPA has indicated that these 17 chemicals were selected either because they are produced or released in high volumes, or because opportunities for reductions in releases of these chemicals to the environment have been identified. EPA is encouraging companies to use pollution prevention strategies to achieve these reductions. As of early 1993, more than 1,000 companies had committed to participating in the program.

Companies that wish to join the 33/50 Program are urged to provide EPA with a letter containing specific commitments to achieving the environmental discharge reduction goals, including a description of the strategies being employed to achieve those reductions. As this information is available to the public, these so-called "commitment" letters will be your best 33/50 information source. Depending upon the extent of the commitment and the quality of the information submitted by the company, you may be able to assess whether source reduction techniques are being used to minimize environmental emissions at the plant. You may also question plant officials during the interview about their progress in meeting the 33/50 goals. More important, you can discuss the role of source reduction in their general 33/50 efforts.

EPA is also collecting individual progress reports from participating companies. These reports, which are available to the public, ask companies to describe:

- Specific pollution prevention activities that reduced or eliminated toxic chemical emissions.

- Steps taken to identify, evaluate, and implement pollution prevention techniques.

- Institutional innovations to foster pollution prevention.

- Economic/environmental benefits and costs of pollution prevention.

- Incentives/barriers affecting pollution prevention.

- Technical assistance needs.

It is important to note that, although plants are strongly encouraged to achieve 33/50 reductions through pollution prevention, a company may also be accredited with these reductions by utilizing end-of-the-pipe waste management techniques such as out-of-process recycling and incineration. The program's focus is on release reduction, rather than strictly on source reduction.

If the plant has not made a commitment to the 33/50 Program, it may be because it does not use or discharge any of the chemicals targeted by the program. However, as the program also encourages companies to reduce discharges of additional chemicals, you can suggest to plant officials that they join their colleagues and competitors in industry and make official reduction commitments to the program.

Section C: Commitments to the US Environmental Protection Agency 33/50 Program

Section C of Worksheet 3 covers information EPA has asked companies to include in their "commitment" letter for the 33/50 Program. If the plant you are studying is part of the 33/50 Program, you should be able to complete Section C by obtaining a copy of the letter or by requesting the information contained in the letter from the 33/50 Administrative Record. If the answers to some of the questions are not available through these sources, you may want to ask for the information during the plant interview.

How to Obtain 33/50 Information

All information submitted to EPA by participating companies is available to the public through the 33/50 Program Administrative Record. Company-specific information will vary and may include location and mailing address, facility identification, 33/50 Program commitments, reduction goals, and company progress reports. Company-specific information is available in various forms, including copies of correspondence, computer-generated lists, and computer files on floppy disks. For more information describing the company-specific information available through the 33/50 Program, or to request such information, contact EPA's Toxic Substances Control Act (TSCA) Hotline:

> TSCA Hotline
> Environmental Assistance Division
> EPA (TS-799)
> 401 M Street, SW
> Washington, DC 20460
> Tel. (202) 554-1404 Monday through Friday, 8 a.m. to 5 p.m.

You can also contact the 33/50 Program itself:

> The 33/50 Program
> EPA TS-799
> 401 M Street, SW
> Washington, DC 20460
> Tel. (202) 260-6907

Another source of information on the company-specific information available through the 33/50 Program is the Emergency Planning and Community Right-to-Know Act (EPCRA) Reporting Center:

> EPCRA Reporting Center
> c/o Computer Based Systems, Inc.
> 4301 N. Fairfax Drive, 6th Floor, Suite 650
> Arlington, VA 22203
> Tel. (703) 816-4445

The Pollution Prevention Information Exchange System (PIES), a free computer bulletin board associated with EPA's Pollution Prevention Information

Clearinghouse, also contains 33/50 information. To learn how to access PIES, call
(703) 821-4800. To access PIES using a personal computer and modem, call (703)
506-1025 (set your communication software to Parity = None; Data Bits = 8; and
Stop Bits = 1). (See Resources at the end of this chapter for more information
about PIES and the Pollution Prevention Information Clearinghouse.)

Bibliographic Reports on Pollution Prevention Options for 33/50 Industries

EPA is creating a series of bibliographic reports for some of the 33/50 industries
that are the largest emitters of the 17 chemicals. Each report provides summaries
of the industrial processes within the category that are primarily responsible for
the release of the chemicals of concern, describes applicable pollution prevention
alternatives, and provides a guided bibliography of other sources that may provide more detailed technical information on potential pollution prevention and recycling methods for the particular industrial processes. The reports are designed
to assist the general public and government officials in learning about pollution
prevention.

As of late 1992, reports had been prepared for four industries: wood manufacturing (SIC Code 25), metal fabrication (SIC Codes 34-38), primary metals
(SIC Code 33), and printing, publishing and allied industries (SIC Code 27). Additional reports covering other industries will be available periodically. You may
ask whether new reports are available when you make your information request
to EPA's Administrative Record or TSCA Hotline. Obtain copies of the reports by
calling PIES.

 ## Worksheet 4 Chemical Hazards and Chemical Storage

Besides information about the toxic or hazardous chemical wastes generated by
a facility, available through the Toxics Release Inventory (TRI), you can obtain
information about the amounts of hazardous chemicals stored at plants and the
hazards associated with these chemicals, as well as the hazards associated with
new chemicals a facility wants to manufacture. Some of this information is available through **State Emergency Response Commissions (SERCs)** and **Local
Emergency Planning Committees (LEPCs).**

Each state is required to appoint a State Emergency Response Comission
which, in turn, is required to divide the state into emergency planning districts and
name a Local Emergency Planning Committee for each district. An LEPC is sup-
posed to be composed of representatives from the following groups: elected officials, police officers, firefighters, health organizations, hospitals, transportation
officials, the media, community groups, and industrial facilities. To reach your
LEPC, call the TRI contact in your state (listed in Appendix A) and ask for the
number of the SERC. That commission can lead you to the appropriate contact

person in your LEPC. Worksheet 4 outlines the data you can obtain when you contact your LEPC or the US Environmental Protection Agency (EPA), and will help you organize the data you collect.

The Emergency Planning and Community Right-to-Know Act (EPCRA) is Title III of the 1986 Superfund Amendments and Reauthorization Act: this law requires industrial facilities to provide the LEPC, the SERC, and the local fire department with **Material Safety Data Sheets, Emergency** and **Hazardous Chemical Inventory** forms, and other information necessary to create a community emergency preparedness plan. Worksheet 4 focuses on Material Safety Data Sheets and chemical inventories. These materials are useful tools in your efforts to assess the health effects and quantities of hazardous materials used or stored at the plant. The information you obtain can help you identify which facilities in your community handle toxic or hazardous substances that may pose a threat to workers and neighbors.

As you will see, these data are very different from the off-site release and transfer information collected under the TRI. This supplemental information will provide you with some perspective on occupational health hazards associated with chemicals and daily and annual average quantities and location of chemicals stored on-site. You will thus be in a good position to ascertain which facilities could benefit most from source reduction efforts.

Information about new chemical substances a plant wants to manufacture is also available through **Premanufacture Notices (PMNs)** companies must submit to the Environmental Protection Agency. You may find this information useful as well.

Section A: Material Safety Data Sheets

SERCs, LEPCs, and local fire departments maintain Material Safety Data Sheets (MSDSs) for facilities required to complete them under the federal Occupational Safety and Health Act's worker right-to-know standard. These data sheets are designed to inform the workers in a plant about the health risks of workplace exposure to specific chemicals used in the plant. A typical MSDS contains product identification; a warning label or statement; information about precautionary measures; first-aid or emergency procedures; occupational safety procedures; data on flammability or reactivity; information on health effects; physical data; and spill, leak, or disposal information. The MSDS requirement applies to all chemicals, chemical mixtures, and products in use today. There are about 70,000 known chemicals, but the number of chemical mixtures is virtually infinite.

Community exposure differs from worker exposure. Workers may be more at risk because they handle the materials in much closer proximity than community residents, often daily. Community residents, however, may be exposed through air and water pollution, or chemical accidents. Communities are exposed to a different degree than workers. For these reasons, although the MSDS information cannot be used directly to understand community risks, it may raise concerns about particular chemicals that you can explore with company representatives.

The MSDSs are available from your Local Emergency Planning Committee or from the companies themselves. As an MSDS is prepared for each chemical, companies have numerous MSDSs; you may wish to request sheets only on certain substances, such as those that are discharged from the plant or are extremely hazardous. It may also be helpful to visit your LEPC headquarters to get a first-hand look at the information.

Section B: Chemical Inventories

All facilities required to prepare an MSDS must also complete an Emergency and Hazardous Chemical Inventory form. These forms list the daily and annual quantities of hazardous substances stored on-site and the location of these substances at the facility. They are prepared for a larger number of chemicals than those covered under the TRI. Thus, the forms are a valuable tool for assessing which chemicals are used and stored on-site at the plant. (See examples of these forms in Appendix D.) Chemical inventory information is available from the LEPC or the state agency.

Section C: New Chemical Substances (Premanufacture Notices)
(optional)

The **Toxic Substances Control Act (TSCA),** enacted by Congress in 1976, established requirements for identifying and controlling potential toxic chemical hazards to human health and the environment. Programs developed under TSCA now gather information about the toxicity of certain chemicals, the extent of human and environmental exposure, and the potential risks of such exposure.

For the purposes of this guide, the most important information available under TSCA is the document that must be completed by the manufacturer or importer of a newly developed chemical substance. This is known as a Premanufacture Notice (PMN). The PMN must be provided to the US Environmental Protection Agency at least 90 days prior to the manufacture or import of the chemical. After EPA reviews the PMN, it is authorized to consent to or deny the manufacture or import of the substance. It is important to note that substances in the following product categories are exempt from TSCA notification: tobacco, nuclear materials, munitions, food additives, drugs, cosmetics, and substances used solely as pesticides.

Obtaining PMN information (or a copy of the individual PMNs) may be useful to your plant research because it provides very specific information about chemicals the plant is manufacturing or importing. In addition to identifying the chemical involved by its Chemical Abstract Services (CAS) number, the PMN notes any permit for discharging the chemical into any body of water through its National Pollutant Discharge Elimination System (NPDES) number. (The NPDES was created under the Federal Water Pollution Control Act, which requires all point source discharges into any body of water to be permitted by EPA or the designated state agency.) A PMN also contains information about how the chemical is used in the plant, occupational exposures, environmental releases, and disposal. An optional section of the PMN calls for pollution prevention information, allowing the facility to report on its efforts to reduce potential risks associated with the manufacturing, processing, use, and disposal of the chemical.

Section C of Worksheet 4 lists information you can obtain through PMNs. Remember that this information is available only on manufacturers and importers of new chemicals. You may find that the company or plant you are researching is not engaged in this activity, or the company itself does not develop new chemicals, or such chemicals are manufactured by the parent company but not used at the subsidiary plant in your neighborhood.

To order a computer printout of the information in a facility's PMN(s), or to obtain a hard copy of the PMN form as it was completed by the manufacturer or importer of the substance, you must send a request in writing to:

Jeralene Green
EPA FOIA Officer
US EPA
A-101
401 M Street, SW
Washington, DC 20460

See Appendix F for pointers on how to write a Freedom of Information Act (FOIA) letter. Be sure to provide the FOIA officer with as much information as possible about the company or plant, such as the name, address, Dun & Bradstreet number (from Worksheet 1), and any other information you know that can help identify the facility. Ask the FOIA officer to direct the Office of Toxic Substances to respond to your request.

Remember to ask the FOIA officer to notify you if your request is especially large or costly so that you have the option to change or cancel the request. For example, if the company or plant you are researching has submitted numerous PMNs in the last several years, you may want to request only the most recent submissions, or perhaps only those submissions which indicated that the chemical substance was particularly hazardous to workers or the environment. If this is the case, an information specialist on the TSCA hotline can help you in reassessing your search. These specialists can also help you understand the information you receive.

The TSCA Assistance Information Service (TSCA Hotline) is available to answer general questions about the PMNs and TSCA; it operates from 8:30 a.m. to 5 p.m. Call (202) 554-1404. You may also write or fax your questions to:

Environmental Assistance Division (TS-799)
Office of Toxic Substances
EPA
401 M Street, SW
Washington, DC 20460
Fax (202) 554-5603

Community Emergency Preparedness, Emergency Releases, and Accidents

Section A: Community Emergency Response Preparedness Plan

In addition to Material Safety Data Sheets and chemical inventories, the federal Emergency Planning and Community Right-to-Know Law (EPCRA) requires facilities to report accidental releases of **extremely hazardous substances** to the Local Emergency Planning Committees (LEPCs). These substances include 360 chemicals on the federal extremely hazardous substance list (created under Section 302 of EPCRA) that could cause serious human health effects from short-term exposures such as accidental air releases. An LEPC can request materials necessary to do emergency planning for any facility using or storing these extremely hazardous substances above threshold amounts. These materials include safety audits, internal hazard assessments, insurance material, source reduction information, and more. Keep this information in mind not only as a source for collecting background data, but also as a mechanism for getting information from an uncooperative company, if necessary. As the community emergency plan contains information on emergency preparedness for all of the facilities covered by the LEPC in the area, you may also find information on other local facilities.

Section B: Accidental and Emergency Releases of "Extremely Hazardous Substances"

The use, production, handling, and storage of hazardous materials can lead to accidents and spills, both in the plant and during the transport of such materials. Learning about the number and severity of such incidents in a particular plant can provide insight into potential threats to the surrounding neighborhood and help you assess a company's commitment to plant and community safety. This information is useful to your plant research because it provides you with a more comprehensive picture of where other problems or hazards currently exist or have existed in the past at the plant.

Facilities must immediately notify the State Emergency Response Commission and the Local Emergency Response Committee in the event of an emergency or accidental release of "extremely hazardous substances." Notification of the release and a written follow-up report must provide details of the incident and its aftermath, including information such as the name of the substance; location of the release; quantity of the substance released; time and duration of the incident; media to which the substance was released — air, water, or land; known or anticipated health risks; necessary medical attention; necessary precautions; and a contact person at the facility. You should be able to obtain this information from the LEPC in your area. Section B of Worksheet 5 lists questions about accidental or emergency releases. Information about accidental releases, spills, and other incidents involving hazardous substances is available through public databases. The most useful of these data sources for the purposes of plant research and promotion of source reduction are described in the discussion of Sections C-E of Worksheet 5.

In some states, the accidental and emergency release information is collected by the state government. Ask your Toxics Release Inventory contact (listed in Appendix A) whether or not this is the case in your state.

Section C: Other Hazardous Substance Releases (Emergency Response Notification System)

The national Emergency Response Notification System (ERNS) database is compiled by the US Environmental Protection Agency from reports to the National Response Center (NRC), the Coast Guard, and EPA regional offices concerning releases of oil and hazardous substances and subsequent responses to such releases. ERNS and the regional databases essentially provide a mechanism for finding information about incidents as they were initially reported to the NRC, which was created by a number of environmental laws to receive initial notification of accidental and emergency releases. The ERNS database includes the initial notification of emergency releases of "extremely hazardous substances" also reported to the LEPC. Although there may be some data overlap if you consult both sources, the ERNS database will cover other hazardous materials not covered by the reports to the LEPC.

When a release is reported to the federal authorities, the individual reporting the release is asked a series of questions about it. Where notification information is verified, more detailed data on the release may be added to ERNS, including information related to response actions. Often, however, the only information found in ERNS is information derived from the initial notification. Once you have obtained the basic notification information and report number(s) from the ERNS database (when you call or write, be sure to ask for both the ERNS and the regional incident report), you can then call the EPA regional office in your state to obtain more specific information about the incident. Each EPA region also maintains its own region-specific database, which is a subset of the national database.

Standard ERNS reports provide a one-page summary of the releases as reported to the federal government. They can be obtained on a floppy disk or as a computer printout, depending upon the number of reports requested. You can ask the ERNS officials to search for specific release dates, specific dischargers, information on specific chemicals, and information on the county, city, or address of the release.

For information about the regional ERNS databases, contact the Freedom of Information Act (FOIA) Office in your region. (FOIA contacts are listed in Appendix F.) Information in the national ERNS database may be obtained through EPA's FOIA office, the FOIA office for your region, or the ERNS office:

ERNS Manager
EPA, OS-210
401 M Street, SW
Washington, DC 20460
Tel. (202) 260-2342

ERNS data is also made available to the public in periodic reports published by EPA's Emergency Response Division. These reports contain summaries of release notifications and can be obtained by calling the RCRA/Superfund Hotline at (800) 424-9346.

You may find that a particular company has made numerous accidental release or spill notifications to the NRC. If so, you may want to question the plant manager about the source of the problem and steps taken to reduce the incidence of accidental releases, spills, etc.

Section D: Transportation Incidents (Hazardous Materials Information System)

The Hazardous Materials Information System (HMIS) is the primary national source of information about transportation spills of hazardous materials. The records go back to 1971. A Hazardous Materials Incident Report is a written report of any hazardous materials unintentionally released during transportation. These reports must be filed by the carrier within 30 days of the occurrence. The report identifies the mode of transportation, name of reporting carrier, shipment information (including the origin and destination of the shipment), results of the incident, hazardous materials involved, nature of the packaging, cause of the failure, and a narrative description of the incident. Specific information or prepared incident reports can be obtained from:

> Information Systems Manager
> Office of Hazardous Materials Planning and Analysis
> DHM-63
> Research and Special Programs Administration
> US Department of Transportation
> 400 Seventh Street, SW
> Washington, DC 20590
> Tel. (202) 366-4555

To obtain information, first call the preceding number and ask to speak to the systems manager. Explain that you are researching a particular plant or plants and that you would like the Department of Transportation to search the database to find out whether that plant was either the origin or destination of any shipment of hazardous materials that was involved in an accident or spill.

Section D of Worksheet 5 lists the information that could be helpful to your research. You essentially want to see the information in the Hazardous Materials Incident Report for any incident involving that plant. Remember to ask before your order is filled about the cost and time involved.

Section E: Accident Prevention (Accidental Release Information Program)

The Chemical Emergency and Preparedness Office (CEPPO) within EPA's Office of Solid Waste and Emergency Response (OSWER) has initiated the Accidental Release Information Program (ARIP) to develop a national database on the causes of chemical accidents and identify methods used to prevent recurrences. ARIP is probably the most unique of all chemical accident databases because it focuses on the root cause of the incident and efforts undertaken by the facility to prevent a recurrance.

A facility is listed in the ARIP database if it has had a chemical release that met at least one of the following criteria:

- The release resulted in a death or injury.

- The release involved 1,000 or more pounds of a hazardous substance with a **reportable quantity (RQ)** of 1, 10, or 100 pounds, or the release involved 10,000 or more pounds of a hazardous substance with an RQ of 1,000 or 5,000 pounds. (The reportable quantity is the minimum amount of the substance released through a discharge or spill that must be reported to the government.)

- The release was the fourth through tenth release in a 12-month period (repeat release).

- The release involved an extremely hazardous substance listed in Section 302 of the Superfund Amendments and Reauthorization Act (SARA).

Facilities with releases that meet any of the above criteria are required to complete an information form containing supplemental accidental release prevention information in addition to the ERNS information on the release (including facility identification, spill location, chemical released, ERNS number, date and time of release, authorities notified, and response action). The supplemental information is divided into three sections: facility profile, hazardous substance release profile, and prevention profile.

The information from this questionnaire is available to the public. Section E of Worksheet 5 contains the information you should request when you contact the Chemical Emergency Prevention and Preparedness Office ; it will help you organize the information you obtain.

To find out whether the plant you are studying is on the ARIP database, begin by calling CEPPO and ask for the person who is currently dealing with ARIP information requests. Explain that you are researching a particular plant and that you wish to request a search for the plant on the ARIP database. If the plant is on the ARIP system, you may then request a printout of all available information on the plant. It is much easier for the EPA official to respond to your request if it is in writing; therefore, call to find out if your plant in on the database, or if you need other immediate information. Submit your request in writing to:

EPA
Chemical Emergency Preparedness and Prevention Office (CEPPO)
401 M Street, SW
Washington, DC 20460
Tel. (202) 260-8600

In your letter, include the name and address (including zip code) of the plant and any other plant identification information you have, such as Dun & Bradstreet number, SIC code, etc.; this will expedite your request. Include your name, address, and phone number so the EPA official can contact you if necessary. Also, be sure to ask that the EPA official notify you of any cost before filling your order.

Worksheet 6 Hazardous Waste Information, Hazardous Waste Reports

Congress enacted the 1976 Resource Conservation and Recovery Act (RCRA), an amendment to the Solid Waste Disposal Act, to address the problem of how to safely manage the enormous volume of municipal and industrial solid waste generated nationwide. It is the primary federal law that covers the generation, transport, storage, treatment, and disposal of hazardous and nonhazardous solid waste.

Four different but related programs have been established under RCRA. The first encourages states to develop comprehensive plans to manage nonhazardous solid wastes, such as household wastes. The second program creates a system for hazardous waste control from generation to disposal. The third regulates underground storage tanks, and the fourth is a medical waste tracking program. With this broad range of programs, much information is available to the public under RCRA.

For the purpose of promoting industrial source reduction, however, this guide focuses solely on the hazardous waste component of RCRA. As defined under RCRA, hazardous waste is any solid waste that is ignitable, corrosive, reactive, or toxic, and may pose a substantial or potential hazard to human health and safety or to the environment. RCRA covers hazardous waste generators, transporters, and treatment, storage, and disposal facilities (TSDFs). This guide focuses on hazardous waste "generators" (those who create waste in the first place), because those facilities are the most likely candidates for source reduction efforts.

Large-quantity hazardous waste generators (facilities that generate more than 2,000 pounds per month of RCRA hazardous wastes) must submit "Hazardous Waste Reports" to their own state or to the US Environmental Protection Agency. If the plant you are studying is a large-quantity generator, you may obtain information about the type and quantity of the hazardous waste, and about any actions the plant is taking to reduce this waste. Worksheet 6 lists the information you can request.

EPA maintains these Hazardous Waste Reports on the Resource Conservation and Recovery Information System (RCRIS) database. The reports are available from state or federal agencies, depending on whether the state has the authority to implement RCRA regulations. As of early 1993, most states manage the RCRA Hazardous Waste Reports. In fact, some of these states require annual submission of the reports and require more information than is required by the federal program.

Hazardous Waste Reports contain, among other things, information on the quantities of hazardous wastes generated and efforts to reduce the generation of those wastes. Since the form asks for information on both waste minimization and source reduction, it is important to remember (as noted in Chapter 2) that waste minimization often includes "end-of-the-pipe" waste management techniques, such as treatment or incineration, whereas source reduction does not. You should remain aware of the distinction as you gather the RCRA Hazardous Waste Report information.

If your RCRA research shows that a large quantity of hazardous wastes is generated and sent off-site for storage, treatment, and disposal, but that little or no waste minimization efforts are reported, you can ask the plant operators whether they have thought about source reduction opportunities applicable to these wastes. If waste minimization or source reduction efforts are reported under RCRA, ask the plant operators for more specific information about the methods used and what prompted their implementation. As RCRA does not cover all wastes generated at the plant, you might also ask plant officials about waste streams not reported under RCRA.

Obtaining Hazardous Waste Report Information

You may request a computer printout of all the information on a hazardous waste generator or TSDF that is reported on the Hazardous Waste Report forms (also called biennial reports). Although the Hazardous Waste Reports contain a number of forms that must be submitted by hazardous waste generators, for source reduction purposes the most important of these forms are as follows:

- Form IC: Identity and Certification (Section VIII contains waste minimization activity)

- Form GM: Waste Generation and Management (Section IV contains waste minimization activity)

  (Sample copies of the IC and GM forms are included in Appendix D.)

As you will see, it is difficult to decipher these forms without guidance. Therefore, when you make your request for copies of the forms or the computer printout of the information, you should also request a copy of the most recent Hazardous Waste Reports Instructions and Forms. *(See Resources at the end of this chapter for the ordering information.)*

You may obtain more specific information about the Hazardous Waste Reports and other general RCRA information by contacting:

Liza Hearnes
EPA, OS-312
401 M Street, SW
Washington, DC 20460
Tel. (202) 260-3393

To obtain a printout of the information on the form filed by the company, however, you must request the information through the Freedom of Information Act officer at the EPA. Send your FOIA request to:

Jeralene Green
EPA FOIA Officer
EPA, A-101
401 M Street, SW
Washington, DC 20460

In your letter to the EPA FOIA officer, be sure to include the name and address of the facility and the facility identification number, if known. (This number is not crucial, but is useful if you have it. You may request the number from the facility itself or it can be found in Section 4.8 of the facility's Toxics Release Inventory Form R, described in the discussion of Worksheet 3.) Indicate that you want a computer printout of the information the facility submitted to EPA in its Hazardous Waste Reports. You may also ask for hard copies of the forms. (You may be referred to your state office for the hard copies; the federal office does not necessarily have them for all states.) Ask the EPA FOIA officer to direct your request to the attention of Liza Hearnes.

Obtaining Other RCRA Information

You may obtain more information about RCRA by calling the RCRA/Superfund Hotline at (800) 424-9346. This number provides general RCRA and Superfund information. (The Superfund deals with hazardous waste problems at inactive or abandoned sites or those resulting from spills that require emergency response.)

Other RCRA information is included in the Resource Conservation and Recovery Information System database: basic handler information; general facility information; process types; permit and closure action events; other permitting information; enforcement and compliance information; corrective action events; the size of on-site hazardous waste management facilities with RCRA permits; the number of RCRA hazardous waste manifests submitted (a transportation safety concern); recent submitters of permit applications to construct on-site hazardous waste management facilities; and more. To learn more about RCRIS and how to obtain general RCRA information, order a copy of the RCRIS reports catalog. (See Resources at the end of this chapter.)

Section A: Contacting Employee Representatives at the Plant

Important issues such as workplace health and safety, jobs, and quality of work are often ignored when companies and environmentalists discuss pollution and environmental regulations. However, the plant workers can be important contributors in your general efforts to promote source reduction. As workers are often at the greatest risk of exposure, they have a substantial stake in reducing toxic chemicals at the source. Workers are often also concerned about toxic pollution in the community outside the plant because they often live nearby. Moreover, workers directly involved in plant operations are often the ones responsible for discovering and implementing source reduction practices.

Before you contact plant officials to set up an interview, it is a good idea to speak with employee representatives. If the plant is unionized, you should determine what union or unions represent the workers and let them know that your group will be discussing source reduction with plant management. As source reduction involves employees, they will most likely appreciate your contacting them prior to meeting with plant officials. Section A of Worksheet 7 contains questions to ask the employee representative.

You may find the union contact or employee representative at the plant by contacting the American Federation of Labor - Congress of Industrial Organizations (AFL-CIO) in your state. You can get the number in the phone book, by calling information, or from the national AFL-CIO headquarters at (202) 637-5000. The AFL-CIO should be able to tell you which of their member unions are at the facility. Officials in the state and federal offices will also be able to answer other general questions. Another potential source for finding out about a union in the plant is the Central Labor Council in your city or county. Consult your local phone book.

The Coalitions for Occupational Safety and Health (COSH), if one exists in your state, are also a good source of union information. The complete list of COSH contacts appears in Appendix E.

Section B: Occupational Safety and Health Data

As discussed in Chapters 1 and 2, hazards associated with the use and production of toxic or hazardous substances are not necessarily limited to environmental emissions and hazardous waste generation. Workers are often routinely exposed to these substances during standard plant operations. Thus, prior to your plant interview, you may want to gather some information about the plant's compliance with existing occupational exposure requirements.

In 1970, Congress enacted the **Occupational Safety and Health Act** "to assure so far as possible every working man and woman in the Nation safe and healthful working conditions and to preserve our human resources." The Occupational Safety and Health Administration (OSHA) was created in the Department of Labor to enforce the law. (See Resources at the end of this chapter for publications about the coverage and implementation of the law.)

Under OSHA, information about some types of in-plant occupational health and safety issues may be available on the plant you are studying. You may request plant-specific information on areas such as plant and workplace inspections, violations, citations issued, penalties assessed, toxic chemical exposures, employee complaints, local and national standards for evaluation of plant compliance, number of employees covered by an inspection, and number of employees at the plant during the inspection. These are just a few examples. If you want to know about company-wide performance, you may request information on health and safety at a particular company's plants nationwide. Section B of Worksheet 7 will help you formulate your OSHA inquiry and organize the information you obtain. You should be interested primarily in information on past or present toxic hazards, such as workplace exposures, unlabeled containers, leaks, spills, or explosions, and large quantities of hazardous materials in storage.

Information under OSHA can be very limited. As of early 1993, national OSHA reform legislation had been drafted and was targeted for introduction in Congress. If passed, this legislation would significantly expand the quantity and quality of public information on occupational hazards.

To obtain the OSHA information that is now available, call the regional office (listed in Appendix E) that serves your state. Ask for the Freedom of Information Act (FOIA) Coordinator in that office. Explain that you are researching a plant(s) in your area (provide the name, address, etc.) and that you are interested in obtaining publicly available information on worker health and safety at that plant. As the availability of information varies from state to state, the FOIA coordinator will be able to tell you whether your information request(s) should be directed to the area (local), regional, or national OSHA office.

The FOIA coordinator should also be able to provide a computer printout of the available information for the facility. This could help you identify areas you would like to explore further. For example, if a computer printout identifies a violation regarding exposures to asbestos, you may want to look into the outcome of the investigation. In some cases, if the inspection or violation took place many years ago, the records may have been destroyed. You will then have reference to the incident only from the original printout, which you may use during the plant interview to question the company about the final outcome. You are not limited to the questions in Section B of Worksheet 7, but it is helpful to begin your OSHA research by requesting this information. Ask the FOIA coordinator what other information is available. You will find that a friendly conversation can lead to your uncovering information that may be useful in your dialogue with union representatives, as well as your interview with plant officials. Once you have talked with the FOIA coordinator and are ready to write your FOIA letter, it is important that you make your request as specific as possible.

Section C: Occupational Health Hazard Information

If employers or authorized employee representatives of a company wish to obtain information about what they believe may be a potential risk to the health and safety of themselves or others at the plant, they may request that the National Institute for Occupational Safety and Health (NIOSH) conduct a Health Hazard Evaluation (HHE) of that facility. A standard HHE involves studying a workplace in a factory, industrial plant, or other worksite. NIOSH conducts such a study to ascertain whether there is an employee health hazard caused by exposure to hazardous substances in the workplace. Not all HHEs require site visits. Sometimes, the employer or employee who requests the HHE simply files certain information with NIOSH, or is referred to the state agency or another resource.

While members of the general public may not request that a Health Hazard Evaluation be conducted, copies of NIOSH HHE final reports are publicly available. Section C of Worksheet 7 lists the information you can obtain from a final report.

To obtain NIOSH information, call the Technical Information Hotline at (800) 356-4674 and ask for copies of any HHE reports for the facility you are investigating. You will need to provide the name and address of the facility. The technical services personnel at this number will transfer you to the HHE section that will give you a facility identification number. (Contact this section directly by calling (513) 841-4252.) You will then be transferred back to the technical services representative, who will take your order for the copy of the HHE report. Like OSHA, NIOSH covers all types of occupational hazards. Be sure to specify that your primary interest lies in toxic exposures or other chemical hazards. NIOSH keeps HHE reports for a limited time. If NIOSH no longer has the report you request, it will provide you with the ordering information necessary to obtain the HHE report from the National Technical Information Service (NTIS). NTIS will charge a small fee for the HHE data.

Worksheet  State Source Reduction Program Analysis

Before conducting the plant interview, you should be aware of state programs that focus on source reduction or waste minimization. Such programs may exist pursuant to state laws, or they may be voluntary state initiatives. Worksheet 8 provides some basic questions to help you ascertain what type of source reduction program exists in your state. (Appendix B contains a list of state pollution prevention contacts.) This will give you an important perspective on whether your local plant is responding to existing legal requirements or a voluntary program. If written materials on your state program are available, such as descriptions of the state program, law, technical assistance, etc., you may request copies and study them yourself. Such materials should provide a fair summary of the focus of the program.

Once you are somewhat familiar with the general guidelines of the program, consider meeting with state agency officials to discuss the program in depth and learn about publicly available data on the plant you are researching. Such a meeting would also give you a chance to develop a relationship with state officials who deal directly with source reduction issues. If the agency official is willing to speak to you on the phone or meet with you, refer to Chapter 6 of this guide for a comprehensive description of what an effective state source reduction program should contain; you may get an opportunity to suggest improvements.

Worksheet 9 — Industrial Trade Associations: Chemical Manufacturers Association Responsible Care Program

Some trade associations representing various industries and companies on the local, state, and national level have created initiatives designed to encourage member companies to engage in more environmentally responsible behavior. Although such programs are not currently widespread, they are beginning to take shape in various industrial and business sectors. Some encourage member companies to make a commitment to follow an environmental code of conduct or join a waste or release reduction program. Four of these programs are outlined in some detail in the US Environmental Protection Agency's *Pollution Prevention 1991, Progress on Reducing Industrial Pollutants* (see Resources at the end of this chapter under "Industrial Trade Associations"). One of these programs is outlined here.

In 1988, the board and member companies of the Chemical Manufacturers Association (CMA) created and adopted an industry-wide initiative called Responsible Care: A Public Commitment. The Community Awareness and Emergency Response (CAER) Code of Management Practices was, according to CMA literature on the program, "designed to assure emergency preparedness and foster community right-to-know." Most important, it demands of each of its members a commitment to openness and dialogue with employees and the community. CMA's community outreach program is designed to assure that member companies that manufacture, process, use, distribute, or store hazardous materials maintain an outreach project to answer the public's questions about environmental releases, pollution prevention, health risks of chemicals (including occupational hazards), and efforts to ensure the safe storage and transportation of hazardous materials.

Under the Responsible Care initiative, chemical companies are encouraged to be open and to share information. However, as CMA cannot dictate the extent to which they are truly "open" and the quality of the information "shared" with the public, the quality of response by member companies is not consistent. This program can, nonetheless, be a useful tool in your efforts to obtain access to plant officials and in-depth information about source reduction and other health and safety measures. In fact, the Responsible Care program may be the mechanism by which facility officials agree to meet with you. Thus, if your plant research indicates that the company is a CMA member, be sure to mention in your letter to plant officials that you are requesting a meeting with them under the Responsible Care program. *(See sample letter to Plant Managers on page 65.)*

The CMA Code of Management Practices under Responsible Care is divided into the following areas: Community Awareness and Emergency Response, Employee Health and Safety, Process Safety, Distribution, Pollution Prevention, and Product Stewardship. Each of these codes contains a statement of purpose, a description of how the particular code relates to the Responsible Care guiding principles, a list of management practices that should be followed to achieve the goal of the code, a section on the company's self-evaluation of progress under the code, and an explanation of how the particular code relates to the other codes.

In the self-evaluation forms, each member company is asked to report annually to the CMA on the stage of implementation of each management practice in the code; companies rate their own performances in each of the aforementioned areas. CMA collects the forms and compiles an industry-wide report that helps identify areas where member companies may need improvement, according to CMA standards.

The Chemical Manufacturers Association does not make these evaluation forms available to the public. You may request a copy of the self-evaluation directly from the company. It is not required to share these documents with you; if it does, it is demonstrating not only a commitment to openness and dialogue with communities, but a willingness to discuss information about progress and the need for improvement in areas that are vital to a successful facility-wide source reduction program. Call for specific information on Responsible Care in your area (such as a list of member facilities or the contact person for a facility), call (800) 624-4321, ext. 20.

General information about the CMA Responsible Care program may be obtained from:

Chemical Manufacturers Association
2501 M Street, NW
Washington, DC 20037
Tel. (202) 887-1100

Worksheet 9 starts with the initial questions to be asked when you contact CMA to obtain information and the name of the facility contact. It then lists some of the questions you may ask the facility contact directly. If you are unable to speak with the facility contact, you may want to ask these questions after you have interviewed plant officials about their source reduction practices.

Preparing for the Plant Interview

Once you have gathered background data, you are ready to contact plant officials for an interview. As public records do not provide a detailed compendium of source reduction activities at individual facilities, an open dialogue with company officials, including the plant manager, is necessary. Further, by creating a positive relationship with company officials, you will be in a better position to encourage them to focus on source reduction strategies. This section guides you through the steps of setting up a meeting and breaking the ice; it also alerts you to some of the arguments company officials may use.

Setting up a meeting

Figure 5-1 summarizes the steps to follow in trying to set up a meeting with the plant manager and in responding if a company refuses to meet with you. As your goal is to establish a cooperative relationship with company officials, it is important that your initial communications with the company, both by phone and by letter, be professional and positive. A sample letter to a plant manager is shown in Figure 5-2 along with attachments (Figures 5-3 and 5-4) to include with the letter: a list of sample questions and a chart asking the plant manager to provide plant-level materials accounting figures for each chemical used at the plant. The plant manager may use the questions and chart to prepare for the meeting and designate the appropriate company employees to meet with your group.

While you are arranging the meeting, you should be designating your interview team. It should include a respected member of the community and a person with technical knowledge. A team of about two or three people is usually an effective size.

Creating a positive relationship

Probably the most effective tool for getting the information you need from company officials on significant source reduction activities will be the development of a cooperative relationship with them. The following pointers can help you establish a positive relationship.

- Do the background research thoroughly and carefully before setting up an appointment with company representatives. Let them know that you have tried to gather as much information as possible before taking their time with a survey.

- Carefully phrase questions about compliance with current environmental laws and permits. Aggressive questioning in this area may place the official in a defensive position, and you will not get the information on source reduction you want.

- If you plan to pursue such areas as EPA's 33/50 Program commitments, accidents and spills, or occupational health and safety, it may be a good idea to explain to the plant officials how these areas relate to source reduction.

- Make clear that you are not interested in a company's trade secret information. You are trying to get a basic source reduction picture, rather than technical details on individual processes.

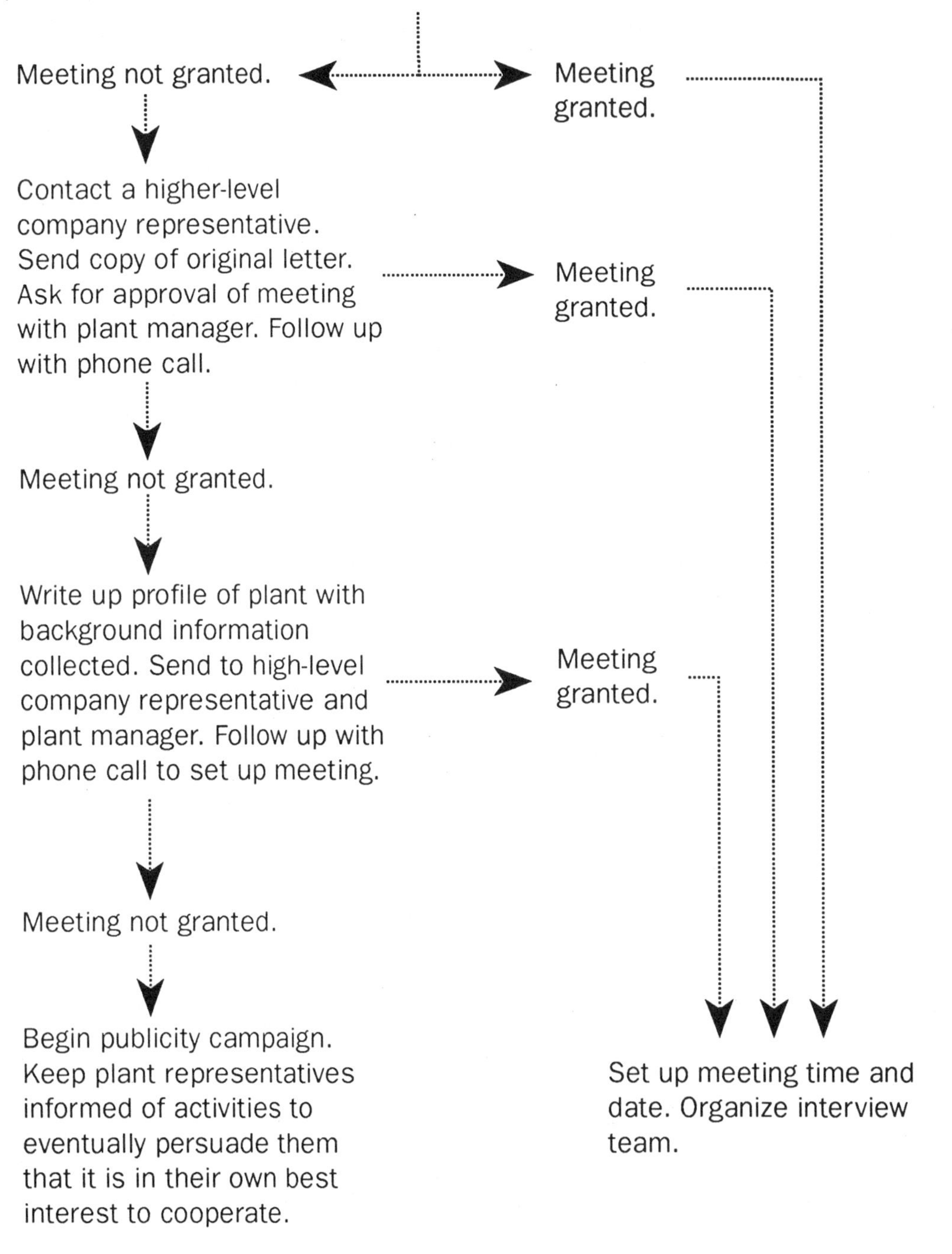

Call plant to find out mailing address and plant manager's name.

Contact and hold meetings with union representatives. (Always keep union representatives informed of your activities.)

Send letter to plant manager (see sample letter).

Follow up with phone call to answer questions and get approval for meeting.

Meeting not granted.

Meeting granted.

Contact a higher-level company representative. Send copy of original letter. Ask for approval of meeting with plant manager. Follow up with phone call.

Meeting granted.

Meeting not granted.

Write up profile of plant with background information collected. Send to high-level company representative and plant manager. Follow up with phone call to set up meeting.

Meeting granted.

Meeting not granted.

Begin publicity campaign. Keep plant representatives informed of activities to eventually persuade them that it is in their own best interest to cooperate.

Set up meeting time and date. Organize interview team.

Group Name
Address
Phone Number

Date

Plant Manager's Name, Title
Plant Name
Address

Dear (Plant Manager's Name):

(Group Name) has begun a program to learn about what generators of toxic or hazardous waste in our region are doing to
reduce, at the source, toxic or hazardous substances, waste
generation, and releases to air, land, and water. We will
be educating interested citizens and others about the extent
of such source reduction activities. We invite you to meet
with us to discuss what source reduction measures have been
taken or are planned at (Name of Plant).

*[NOTE: If the company is a CMA member, you may want to men-
tion that you are requesting a meeting under the Responsible
Care program by adding: We understand that under the Chemical Manufacturer Association's Responsible Care program, one
pledge in the "Guiding Principles" is "to recognize and respond to community concerns about chemicals and our operations." We are confident that, as a CMA member, you recognize and abide by the principles of the program. Thus, we
invite you to meet with us to discuss what source reduction
measures have been taken or planned at (Name of Plant.)]*

Source reduction is defined in the federal Pollution Prevention Act of 1990 as "any practice which (i) reduces the
amount of any hazardous substance, pollutant, or contaminant
entering any waste stream or otherwise released into the environment (including fugitive emissions) prior to recycling,
treatment, or disposal; and (ii) reduces the hazards to public health and the environment associated with the release
of such substances, pollutants, or contaminants. The term
includes equipment or technology modifications, process or
procedure modifications, reformulation or redesign of products, substitution of raw materials, and improvements in
housekeeping, maintenance, training, or inventory control.

(Over, please)

The term 'source reduction' does not include any practice
which alters the physical, chemical, or biological characteristics or the volume of a hazardous substance, pollutant, or
contaminant through a process or activity which itself is not
integral to and necessary for the production of a product or
the providing of a service." *[NOTE: If your state has a
source reduction, toxics use reduction, or pollution prevention definition, you may want to use it instead of the federal definition.]*

Numerous case studies by government, private, and industry
groups have shown that such source reduction measures often
save companies money.

We are not concerned with proprietary information but with
the overall picture of source reduction programs and achievements at your facility. The attached list will give you an
idea of some of the kinds of questions we will be asking. In
addition, we are requesting that you complete the enclosed
chart with facility-level materials accounting information
for each toxic or hazardous substance used at your plant on a
chemical constituent basis. This information will help us
with our assessment of the quantities of substances in the
plant. This should also help you prepare for our meeting and
designate the appropriate plant officials to meet with us. We
also ask that you notify the union(s) in your plant and invite them to participate in our meeting.

In case you are not familiar with (Group Name), we are a
group of citizens from the (name of area) area who work together to reduce toxic risks in our communities and improve
environmental quality in our region. Our previous activities
have included (some recent activities).

We hope that with your cooperation, and that of other hazardous substance users and waste generators in the region, we
can move toward reducing the region's overall toxics burden.
We will call you within the next few weeks to follow up on
this letter and set a date for our meeting.

Sincerely,

cc: Union representatives

Figure 5-3: Enclosure for Sample Letter to Plant Manager

(Include the following questions as an enclosure with your letter to the plant managers.)

Sample Questions for Plant Interview

1. **Responsibility for source reduction policy**
   - Does your company/plant have an official, written policy on source reduction and toxic waste management?
   - Which division/office/person oversees the implementation of this policy?

2. **Corporate mechanisms to implement source reduction**
   - What are the elements of your plant's source reduction program? For example, is there any source reduction training and incentives program for process operators?
   - Does the plant collect facility-level and process-level data? What is the scope of this information collection?
   - Is there a full cost accounting system for toxic or hazardous waste?
   - Who, within the plant, is responsible for reducing toxics and the generation of hazardous waste?
   - What actions have been taken to reduce toxic or hazardous waste generated by the plant? For each action, what was reduced, how was the source reduction opportunity identified, what factors influenced the choice of source reduction techniques used, and how much reduction was achieved?

(Include the following questions as an enclosure with your letter to the plant managers.)

<table>
<tr><td colspan="2" align="center">FACILITY LEVEL THROUGHPUT DATA</td></tr>
<tr><td>Chemical Name: ___________________________________</td><td>CAS Number: __________-____-_____</td></tr>
<tr><td>Inputs</td><td>Quantity (lbs)</td></tr>
<tr><td>Starting Inventory</td><td></td></tr>
<tr><td>Quantity Produced On-Site</td><td></td></tr>
<tr><td>Quantity Brought On-Site</td><td></td></tr>
<tr><td>Outputs</td><td>Quantity (lbs)</td></tr>
<tr><td>Quantity Consumed On-Site</td><td></td></tr>
<tr><td>Quanity Shipped Off-Site as (or in) Product</td><td></td></tr>
<tr><td>Quantity Otherwise Generated</td><td></td></tr>
<tr><td>Ending Inventory of Substance</td><td></td></tr>
</table>

(Photocopy this chart for inclusion in your letter to plant managers.)

Of course, if the company attitude toward citizen inquiry prevents the creation of a positive relationship, you might have to consider other tactics, such as publicity.

Arguments to Anticipate

As you conduct your interview, company representatives may try to explain why the company does not have an adequate source reduction program. Several arguments are frequently offered; with enough background information, you will be able to avoid being put off by them.

"We're in compliance with the law."

Companies traditionally evaluate their environmental management program according to whether it is successful in preventing permit violations or enforcement actions brought by regulators or citizens groups. Many companies have designed audits to assure compliance with laws and regulations, and the person you are interviewing may attempt to share this information with you. These compliance audits are a welcome sign that companies are taking their legal responsibilities seriously. However, they are not a substitute for a source reduction program, because existing regulations are largely focused on end-of-the-pipe control strategies and affect only a fraction of the pollutants generated by industry.

- Many chemical substances remain unregulated. For example, the new Clean Air Act Amendments cover only 189 substances out of the thousands of possibly toxic air pollutants.

- Many pollution sources are unregulated. For example, the Clean Air Act and the Clean Water Act regulate only point sources of contamination (specific, planned emission points such as smokestacks and discharge pipes). They do not capture the large volume of contaminants released as fugitive air emissions (from unintentional release points such as leaky valves or evaporation tanks) or as runoff from plant grounds.

- Even the existing regulations allow a certain amount of pollutants (defined in the plant's permits) to be emitted into the environment. The combined emissions from several plants can mean that a large volume of pollutants are still being released.

In general, a company's source reduction program should not be limited by whatever minimum is needed to comply with the existing laws. You might ask the company representatives if they believe that more regulations are needed to get the industry to make additional reductions in the use of toxic or hazardous chemicals and the generation of toxic or hazardous waste (and if so, what kind). If not, the industry will have to demonstrate clearly how it is moving beyond the limits of existing regulations.

"The pollutants released from this plant pose a very low risk to local citizens."

You are likely to hear that the plant's toxic or hazardous releases pose a relatively low risk to the surrounding environment. The company representative may point out that the plant is already required to comply with occupational exposure limits to protect the health of workers (sometimes called "threshold limit values"), and that exposures to inhabitants of the surrounding neighborhood are many times lower than these occupational limits. While it is essential that companies protect the health of their workers, doing so may not be sufficient to protect neighboring communities or the environment.

- Much of the data about the health effects of exposure to chemical contaminants is speculative. For example, exposure limits may be based on animal studies and limited by unprovable assumptions about the relationship between the dose to which an individual is exposed and the likely health effects. These studies are usually limited to individual chemicals and do not consider the multiplier effects that may occur when several chemicals interact in the environment. Further, the threshold limit values developed by the government are based on workday, not round-the-clock, exposures.

- Concern about chemical contaminants should not be limited to human health. In many cases, although the health risks may be relatively low, more serious effects may be felt by the natural environment, through ecological damage to rivers, lakes, and soils, to cite a few examples. The Great Lakes provides a classic example of an ecosystem where contamination by many different chemicals from many different sources has produced serious damage. For example, the International Joint Commission (IJC), a government body with responsibility for environmental quality of the Great Lakes, reported in 1992: "The principal problem is the presence and impact of persistent toxic substances on all sectors of the ecosystem... Attempts to regulate persistent toxic substances have not resulted in an efficient or successful set of programs." The IJC report concluded: "It is time to ask whether we really want to manage persistent toxic substances after they have been produced, or whether or not we want to eliminate and prevent their existence in the ecosystem in the first place."[1]

Given these risks and uncertainties, and the potential economic benefits of source reduction, the more prudent course is to take preventive actions now that might prevent problems from arising in the future.

"Source reduction at this plant is technically impossible, or too expensive."

As discussed in Chapter 3, many of the opportunities for cutting the generation of potentially toxic or hazardous wastes involve no-technology or low-technology options, such as improving maintenance procedures. For example, a plant in New Jersey has instituted a plant-wide policy whereby all quality control and raw material samples are sent back to be reused in production. This practice reduces hazardous and nonhazardous solid waste by 3,000 pounds per year and results in cost savings of $20,000 per year.

Evaluations by INFORM, the congressional Office of Technology Assessment (OTA), and private industry have shown that some source reduction opportunities may be found in almost every case where a thorough process-level source reduction inventory has been undertaken. Accordingly, if a company representative cites economic or technical barriers, ask how this was determined:

- Have you conducted a thorough source reduction inventory for each process used at this facility?

- If you encountered cost barriers, do you use cost accounting principles that reflect the costs pollution adds to a particular process?

- If you encountered technical barriers, have you sought technical assistance from the state or local government, or your trade association? If so, please describe the source of technical assistance and the type of assistance received (for example, an information packet sent by mail, on-site technical assistance, or referral to an equipment vendor)?

- In your evaluation of source reduction options, did you consider no-technology or low-technology choices as well as more significant changes to production or design? If so, please describe.

"I can't share that with you because it's a trade secret."

Companies are often reluctant to share the details of a specific technological process out of concern that it may provide some advantage to a competitor. While this concern is often well-founded, at times it is the result of a general tendency to play it safe by throwing the "trade-secret blanket" over large quantities of relatively harmless information. If you are faced with this claim, make the following points:

- You are not asking for a detailed description of a "proprietary" process. The source reduction accomplishments chart (Worksheet 11) asks only for a general, categorical description of the type of process used and the source reduction savings that resulted.

- Many companies — such as Dow, Monsanto, General Electric, and Union Carbide — have already provided this kind of data in response to a survey by Tufts University's Center for Environmental Management.

- 1989 Toxics Release Inventory data indicate that more than 2,400 industrial facilities (or 11 percent of all TRI reporters) have already made such information available by filling out the optional source reduction form attached to their TRI form. In 1990, the TRI was expanded in the federal Pollution Prevention Act to include mandatory source reduction data reporting.

- Other companies have provided much more detailed information about specific source reduction processes and the results they achieved. This information has been compiled by North Carolina's Pollution Prevention Pays Program and the US Environmental Protection Agency's 33/50 Program.

"We have already achieved all the source reduction possible."

Research by OTA, INFORM, and others has concluded that most plants have not systematically looked for opportunities to reduce waste generation. Does this company have a systematic approach in place? If the company has not undertaken any of the organizational strategies discussed in Chapter 4, the chance that it has reached its source reduction potential is close to zero. Ask the company representative what success the company has achieved and how it seeks new opportunities.

Research by INFORM in *Environmental Dividends: Cutting More Chemical Wastes* indicates that a great number of opportunities to reduce waste at the source are available for those industrial leaders who seek them at their facilities. Even plants that have already made significant reductions in certain of their waste streams continue to find low-cost, low-technology options to convert raw materials into valuable products more efficiently. A breakdown of the source reduction activities at the plants studied in *Environmental Dividends,* reported by the year they were first implemented, shows that plants — including those that already achieved significant reductions — found more source reduction activities each year in the second half of the 1980s than in any previous year.

"We measure source reduction by waste streams, not specific chemicals."

The Resource Conservation and Recovery Act regulates solid hazardous waste as "waste streams" made up of a number of chemical components. For example, RCRA K104 wastes are "combined wastewater streams generated from nitrobenzene/aniline production."

However, source reduction opportunities and achievements are most effectively evaluated by conducting a chemical materials accounting (described in Chapter 4) for each process in a plant. The best way to conduct such a materials accounting is on an individual chemical basis, because complex mixtures can be altered substantially and become difficult to trace as they pass through the many possible steps in a process. For example, if the more volatile portions of a complex solvent mixture evaporate in a process, two completely different streams are created (the volatile compounds in the air and the residual liquids). Thus, materials accounting measurements become overly difficult if the units being measured are not specific component chemicals.

Of course, you will have to accept waste stream data if that is all the company has, but be sure to ask the company representative about the usefulness of this kind of information in measuring source reduction:

- Are you measuring air pollution and certain types of wastewater discharges that are not released as a "waste stream," or is your source reduction program geared only toward solid wastes regulated under RCRA? (If the latter, it is not a true multimedia source reduction program.)

- How do you know whether you have reduced those toxic components of a given waste stream that are the cause for concern? Have you simply "dewatered" the waste stream (that is, reduced its volume by removing water) or taken out some other harmless component?

- If the company is already reporting chemical-specific emissions for the TRI, why is it not possible to measure source reduction on a chemical-specific basis?

"Our TRI data show that we have reduced our emissions since the 1987, 1988, 1989, or 1990 reporting years."

This type of claim is commonly made, and in fact the TRI emissions numbers in many cases do show some reductions. However, these "paper" reductions in TRI figures most often cannot necessarily be attributed to true source reduction activity. A National Wildlife Federation study (see Resources at the end of this chapter) has revealed that most of the largest decreases in toxic emissions have resulted from changes in reporting requirements (such as chemical reporting thresholds and chemical delistings); analytical methods (calculation changes); and production level variations (volume reductions, such as lowered production) — not from source reduction, waste minimization, recycling, or treatment. Although there are some good examples of emissions decreases due to genuine source reduction practices, you are still advised to request specific information on how the reductions were achieved.

Interviewing Plant Representatives

Your interview with plant representatives is designed to help you learn about the company's source reduction activities. You will notice, however, that many of the questions on the plant visit worksheets (Worksheets 10 and 11) focus on institutional structures that encourage source reduction, not technical assessments of the company's source reduction program. As discussed in Chapter 1, unless you have an extensive technical background, the most effective role you can play is to open a dialogue on the institutional questions. If you succeed, you will have the whole company seeking individual source reduction options. In essence, you will be measuring the corporate commitment to source reduction.

Note that the contents of Worksheets 10 and 11 mirror the format used by INFORM in researching 29 organic chemical plants for its 1992 report, *Environmental Dividends: Cutting More Chemical Wastes.*

As discussed in Chapter 3, source reduction strategies range from simple operations, equipment, and process changes to more sophisticated changes in chemical use or product design. Using these strategies effectively requires fundamental changes in the corporate culture. A commitment by corporate management to source reduction should be reflected in the policies and programs at the plant. INFORM's case study research, and that of others, suggests that such programs should include five key features:

1. A written environmental policy favoring source reduction.

2. A high-level official to oversee source reduction progress (including joint leadership from both the environmental and non-environmental plant managers).

3. A data collection system for identifying sources of waste.

4. A system for accounting, back to its source, the full cost of generating waste.

5. Employee involvement, training, and reward programs.

Although some successes may be achieved without a high-level, continuing commitment to source reduction, many other opportunities will go unnoticed by those with the authority to make changes. The questions on Worksheet 10 will help you assess the extent to which the company operating the plant you are studying has instituted these key elements of a source reduction program.

Section A: Establishing a Source Reduction Policy

The corporate source reduction policy applied to the plant you are studying should be multimedia (applying to air, water, and land), to avoid the "toxics shell game," and include a clear definition of source reduction. In discussing source reduction with corporate officials, it is important to clarify at the outset what activities they understand source reduction to encompass. (See Box 2-1: Why "Waste Minimization" and "Pollution Prevention" Are Not Necessarily "Source Reduction.") Earlier surveys have shown that many companies have understood source reduction to include almost any waste management activity except land disposal (incineration, for example).

Section B: Creating Company Leadership

High-level leadership is essential for an effective source reduction program. INFORM's research shows that combined leadership from both the environmental department and the plant manager or other nonenvironmental department results in greater source reduction progress overall than leadership from just one of these areas. This and other elements of an effective source reduction program are discussed in Chapter 4.

Section C: Involving Employees

INFORM's research shows that many source reduction opportunities are identified by production personnel; plants with employee involvement programs find more source reduction opportunities than those without them. Employee involvement programs include source reduction training, solicitation of ideas, and recognition and rewards for accomplishments initiated by employees. Ideas should be routinely solicited from all employees: managers as well as plant floor operators. Multidisciplinary teams are similarly useful for identifying source reduction opportunities, because they bring together key people from all affected departments (such as research and development, process engineering, production, marketing, environment, and health and safety).

Section D: Collecting Information

A systematic program for tracking quantities of toxic chemicals as they are used, produced, or wasted in industrial processes throughout a plant (from storage areas to manufacturing, processing, and off-site shipment) is essential to a source reduction program because it provides the information needed to identify source reduction opportunities and measure source reduction achievements over time. Without a system for measuring how much waste generation has been prevented, it is impossible to determine the effectiveness of specific source reduction methods. Thus, the existence of a systematic data-gathering program in a company or plant reflects a serious commitment to long-term source reduction.

The optional **materials balance** data questions apply to plants that use a more detailed form of materials tracking called a "materials balance" or an "engineering mass balance." This is a more rigorous calculation than a "materials accounting" described in Chapter 4. A materials balance itemizes chemical inputs and outputs at individual processes. It aims to account for every pound of a chemical that is (a) shipped to the process, (b) created or destroyed in the process, (c) delivered as a product from the process, and (d) wasted (whether it is an air, water, or solid waste). If the amount of wastes identified does not equal the difference between the amount of the chemical entering or being created in the process and the amount of the chemical leaving or being consumed in the process, then other sources of waste must exist.

For example, if a particular process receives 3 million pounds of a particular chemical as input, and if 2 million pounds of that chemical appear in the product or are consumed in the process, then the difference of 1 million pounds must be released as waste. If the plant manager measures only 600,000 pounds of waste at the end of the pipe, it would be necessary to find other ways waste was being released during the process. Plant management addresses such discrepancies by rechecking measurements, measuring additional areas that may be sources of waste, and preparing task forces to identify likely reasons for the discrepancies.

It is important for two reasons to know the level and nature of uncertainty in the measurement techniques used for the materials balance. First, for process-level source reduction inventories that seem to balance, the level of uncertainty may indicate that certain amounts of material may have been missed. On the other hand, if such an exercise does not balance, but the difference is within the level of uncertainty, the difference might be due to inherent measurement inaccuracies and not necessarily indicate a release into the environment.

For instance, in the previous example, a facility inputs 3 million pounds of a chemical raw material for a particular process. Measurement techniques for such large quantities are typically accurate to within about 5 percent. In this case, that translates to an accuracy of +/- 150,000 pounds. Therefore, materials balance calculations might not detect releases of up to 150,000 pounds of this chemical from the process. For those plants with materials balance practices, be sure to ask what the level of uncertainty is in their measurement techniques, and where discrepancies exist. Those areas may be good targets for source reduction.

Statistical process controls (SPCs) are another measurement technique used by plant operators. INFORM's research shows that plants using SPCs find many source reduction opportunities for the chemicals and processes to which they have been applied. SPCs are systematic techniques to determine the quantitative relationship between process parameters — such as temperature or reaction time — and product quality, chemical use, or waste generation.

Section E: Using the Collected Information

The questions in this section are aimed at clarifying how and by whom the materials tracking data are used as a tool for identifying opportunities for source reduction and/or monitoring progress.

Section F: Assigning the Full Cost of Waste Back to the Source

When the costs associated with waste generation are accounted back to the process in which the waste was generated, the financial benefits of reducing such waste can be more clearly defined, thus motivating action. As noted in Chapter 4, INFORM's research has shown that those *Environmental Dividends* study plants that assigned these costs back to the process implemented an average of three times as many source reduction activities as those plants with no such accounting system.

The types of costs that might be accounted back to the source include materials (e.g., the costs of wasted starting materials and lost products); waste handling (e.g., capital and operational expenses for on-site recycling, treatment, storage, or disposal facilities); transportation and other expenses for wastes sent off-site; regulatory compliance (e.g., staff time, other resources used for compliance with environmental regulations); insurance premiums; future liabilities from accidents, worker illness, or waste site clean-up; and public/customer relations dealing with toxic chemical use or waste issues.

Section G: Preparing Source Reduction Plans
and Progress Reports

This section asks for information on the extent to which the plant prepares reports on source reduction progress and states goals and timetables for future projects. The type of information that could be collected includes the type of action taken, its capital costs, potential savings, payback periods, and research and development and implementation time requirements.

Worksheet 11 Source Reduction Activity Chart

After finding out about the plant's institutional structure and its program for collecting toxic or hazardous substance-related data, both of which give you information about the company's commitment to source reduction, you will still need to document the company's actual source reduction accomplishments. Ask the company representative to help you fill out the achievements profile (Worksheet 11) for any source reduction activity the plant has implemented. The worksheet can also be used to gather information about the company's future source reduction plans. (An example of how to fill out the worksheets is shown.)

Conclusions and Follow-up

Once you have gathered background information from all available sources and learned as much as possible from your meeting with company representatives, you will be able to draw conclusions about the plant you have studied. You will know whether:

- It is using or producing toxic or hazardous substances.

- It is generating toxic or hazardous waste.

- It is in compliance with existing environmental laws.

- It is in compliance with occupational health and safety requirements regarding toxic exposures.

- It has had accidents or spills of toxic or hazardous substances.

- It has made commitments to industry or governmental initiatives that promote source reduction.

- It has made a policy and staff commitment to source reduction.

- It has established a systematic program to identify source reduction opportunities and associated cost savings.

- It has assessed source reduction progress.

- It has outlined source reduction plans for the future.

What Next?

Use the data to identify opportunities for the plant to pursue source reduction practices more effectively. While the primary purpose of this guide is to help you obtain this information, a brief summary of further activities you could undertake follows. For more details and practical suggestions, see guidebooks describing strategies for citizen organizing published by the Citizens Clearinghouse for Hazardous Waste, Greenpeace, and the National Toxics Campaign. The National Toxics Campaign and the Working Group on Community Right-to-Know also have published various case studies of citizen campaigns *(see "Citizens Organizing Tools" in the Resources section at the end of this chapter).*

Whether you assess the plant you studied to be a shining star or a worst-case example, publicizing the findings through local news media or other outreach campaigns may prove effective in promoting more source reduction initiatives in your area. If the study plant has established a successful, systematic source reduction program, public recognition might encourage managers of other plants in your area to follow its example. On the other hand, if the study plant has made few efforts to identify and pursue source reduction opportunities, a clearly presented comparison of the steps outlined in this guide with the steps lacking at the study plant could spur the company, and others like it, to take more aggressive action.

Your group should be acquainted with the local media that may be interested in your progress with the plant. Local environmental, community, or labor organizations may be sources of advice on how to conduct a public campaign. There are numerous examples of both positively and negatively focused citizen campaigns that show how an informed group of concerned citizens can have a significant impact on plants that use and release toxic or hazardous substances.

Your findings might be the basis for a forum promoting discussion among community leaders, local government officials, and plant representatives about specific steps that could be taken to encourage and document source reduction activity and progress. Such a forum occurred in March 1988, in West Virginia's Kanawha Valley. The workshop brought together industry leaders, government officials, and citizens to discuss their concerns regarding toxic releases into the local environment. The workshop resulted in continuing dialogue among these groups and efforts to create a "scorecard" that local plants could complete to convey their progress to local citizens groups.

The findings may also be useful in organized legal actions directed at facilities found to have poor records. You will find the aforementioned guidebooks particularly useful in this area.

As an additional step, your group could use the RISK*ASSISTANT software package to evaluate the exposures to and risks from environmental contamination in your area. Call Thistle Publishing at (703) 684-5203 for information about the computer program. Remember, however, that your efforts should be focused on promoting an effective source reduction program at the local facility, because preventing the creation of waste in the first place is the best way to reduce environmental and public health risks, whether known or unknown.

Resources

Permit Compliance and Applications

US Environmental Protection Agency Office of Water Enforcement and Permits (EN-355). *A Primer on the Office of Water Enforcement and Permits and Its Programs.* March 1990. (Available from this EPA office, 401 M Street, SW, Washington DC 20460. (202) 260-5682.)

Toxics Release Inventory and Pollution Prevention

Citizen Action. *Poisons in Our Neighborhoods.* (Citizen Action, 1120 19th Street, NW, Washington, DC, 20036. (202) 775-1580. $50 for each state guide, $75 for a national listing.).

A comprehensive state-by-state analysis of Toxics Release Inventory data.

INFORM. *Toward A More Informed Public: Recommendations for Improving the Toxics Release Inventory* 1991. (Jacqueline B. Courteau and Nancy Lilienthal). See p. 198 for ordering and discount information. ($10)

Contains 24 recommendations for improving the TRI program.

National Wildlife Federation. *Phantom Reductions: Tracking Toxic Trends* (Gerald V. Poje and Daniel M. Horowitz). (National Wildlife Federation, 1400 16th Street, NW, Washington, DC 20036. (800)-822-9919. $10.)

Examines toxic chemical releases from 29 manufacturers between 1987 and 1988. This report concluded that the largest decreases in emissions between the two years were due to changes in reporting requirements, analytical methods, and production volume, and not from source reduction or pollution control.

The following two publications are available from OMB Watch at 1731 Connecticut Avenue, NW, Washington, DC 20009-1146. (202) 234-8584.

OMB Watch. *Community Right-to-Know: A New Tool for Pollution Prevention.* (Free)

OMB Watch. *Using Community Right-to-Know: A Guide to a New Federal Law.* (Free)

Tufts University Center for Environmental Management. *Managing Chemical Risks: Corporate Response to SARA Title III.* (Center for Environmental Management, Tufts University, Curtis Hall, 474 Boston Avenue, Medford, MA 02155. (617) 381-3486.)

An analysis of how SARA Title III has influenced corporate behavior.

Each of the EPA resources listed below is available through the Emergency Planning and Community Right-to-Know Information Hotline unless otherwise indicated. Toll-free (800) 535-0202; in Washington DC (202) 479-2449.

US Environmental Protection Agency Office of Pollution Prevention and Toxics, and Office of Environmental Engineering and Technology. *Pollution Prevention Resources and Training Opportunities in 1992*. EPA 560/8-92/002. January 1992.

A comprehensive resource on pollution prevention.

US Environmental Protection Agency Office of Toxic Substances. *A Guide to the Toxic Chemical Release Inventory Form R*. EPA 560/4-88-010. September 1988.

Ask for the updated version of this guide. It should include any changes to Form R such as new source reduction and recycling data recently added to the form via the 1990 Pollution Prevention Act.

US Environmental Protection Agency Office of Toxic Substances. *Toxic Chemical Release Inventory Questions and Answers*. EPA 560/4-91-003. Revised 1990.

US Environmental Protection Agency Office of Toxic Substances. *Common Synonyms for Chemicals Listed under Section 313 of the Emergency Planning and Community Right-to-Know Act*. January 1988.

Provides synonyms for chemicals listed in the Toxics Release Inventory, enabling you to look them up in the database.

US Environmental Protection Agency Office of Pollution Prevention and Toxics. *Toxic Chemical Release Inventory Reporting Form R and Instructions*. Revised 1991. EPA 700-K-92-002. May 1992.

A summary of changes to Form R, a copy of the new form, and directions for completing it.

US Environmental Protection Agency. *Chemicals In Your Community: A Guide to the Emergency Planning and Community Right-to-Know Act*. 1988.

US EPA Pollution Prevention Information Clearinghouse and the Pollution Prevention Information Exchange System

EPA has developed a clearinghouse and database that cover pollution prevention (source reduction) information. These are known as the Pollution Prevention Information Clearinghouse (PPIC) and the Pollution Prevention Information Exchange System (PIES).

Information distributed in the PPIC includes general information packets and technical information for certain specific industries, such as metal finishing and printing. When you request an information packet, you are added to the mailing list for "Pollution Prevention News" and the PPIC general announcements. You can reach PPIC at:

> Pollution Prevention Information Clearinghouse
> c/o SAIC
> 7600-A Leesburg Pike, Room 369
> Falls Church, VA 22043

Information on the PIES computer database includes an interactive message center; bulletins containing pollution prevention news and feature articles; a calendar of upcoming events and conferences; literature search databases on case studies and general documents on pollution prevention; a legislative tracking system; summaries of state and federal pollution prevention programs; and a directory of pollution prevention experts.

To obtain PPIC information by phone, or for information on how to access PIES, call:

> RCRA/Superfund Hotline (800) 424-9346 or
> PPIC Technical Assistance (703) 821-4800
> Fax (703) 442-0584

US EPA 33/50 Program

The following 33/50 publications are available free of charge by calling or writing the TSCA Environmental Assistance Division, US EPA (TS-799), 401 M Street, SW, Washington, DC 20460. (202) 554-1404. They are also available through the Pollution Prevention Information Clearinghouse (see previous section).

EPA's 33/50 Program: A Progress Report, Reducing Toxic Risks Through Voluntary Action. July 1991.

EPA's 33/50 Program: Second Progress Report, Reducing Risks Through Voluntary Action. February 1992.

> *These two 33/50 Progress Reports were available as of early 1993; more recent reports may become available.*

US Environmental Protection Agency Special Projects Office (TS-792A), Office of Toxic Substances. *The 33/50 Program: Forging an Alliance For Pollution Prevention.* July 1991, 2nd Edition.

"Companies Committed to the 33/50 Program By State." February 1992.

> *Parent companies that have expressed their intention to participate in the 33/50 Program. They are listed by the state where the parent company headquarters are located, which may differ from the location of companies' subsidiary facilities. Ask for the most recent list.*

Also ask for any other fact sheets, flyers, or pamphlets that are sent to companies or are available to the public, including the March 1992 sheet entitled, "Suggested Language for the 33/50 Program Company Goals."

Chemical Hazards and Chemical Storage

Benjamin A. Goldman. *The Truth About Where You Live, An Atlas for Action on Toxins and Mortality.* (In bookstores, or call (800) 733-3000 to order. ISBN Number 812-918-3. $17 paperback.)

Contains maps that illustrate which US communities suffer disproportionately from environmental contamination, disease, and death.

Nicholas A. Ashford and Claudia S. Miller. *Chemical Exposures: Low Levels and High Stakes.* Van Nostrand Reinhold, 1989. (Van Nostrand Reinhold, 7625 Empire Drive, Florence, KY, 41042. Credit card orders: (800) 842-3636. $19.95 plus state sales tax.)

New Jersey Department of Health. *Hazardous Substance Fact Sheets.* (NJDOH, Right-to-Know Program, CN 368, Trenton, New Jersey 08625-0368. (609) 984-2202.)

See Appendix C for more information on the Hazardous Substance Fact Sheets.

The following EPA documents are available through the Emergency Planning and Community Right-to-Know Information Hotline. Toll-free (800) 535-0202; in Washington DC (202) 479-2449.

US Environmental Protection Agency Office of Pesticides and Toxic Substances. *Toxic and Hazardous Chemicals, Title III and Communities, An Outreach Manual for Community Groups.* EPA 56-1-89-002. September 1989.

US Environmental Protection Agency. *Chemicals In Your Community: A Guide to the Emergency Planning and Community Right-to-Know Act.* 1988. (Emergency Planning and Community Right-to-Know Information, OS 120, 401 M Street, SW, Washington, DC 20460.)

New Chemical Substances

The following EPA documents are available from the EPA Office of Pollution Prevention and Toxics, TSCA Assistance Office (TS-799), Washington, DC 20460, or call the TSCA Hotline at (202) 554-1404.

US Environmental Protection Agency Office of Toxic Substances. *A Layman's Guide to the Toxic Substances Control Act.* EPA 560/1-87-001. June 1987.

This pamphlet describes TSCA in detail: the scope of the law, programs created, disclosure of data, state programs, etc. It is a source for assessing the potential usefulness of the TSCA information.

United States Environmental Protection Agency Office of Toxic Substances. *New Chemicals Program.* EPA 560/1-91-005. November 1991.

This booklet provides an in-depth explanation of how the Premanufacture Notice and significant new use process works under TSCA.

Note: When ordering these documents, ask for a blank copy of the Premanufacture Notice (PMN) and instructions, see the actual form completed by the manufacturer. This will help you formulate other specific questions about the plant you are studying and aid in your understanding of PMNs in general.

Emergency Preparedness, Emergency Releases, and Accidents

Community Emergency Response Plan, Emergency Releases of "Extremely Hazardous Substances"

New Jersey Department of Environmental Protection and Energy, and the Association of New Jersey Environmental Commissions (ANJEC). *Community Right-to-Know for Health and Safety, An Outreach Handbook for Local Emergency Planning Committees.* (ANJEC, PO Box 157, Mendham, NJ 07945. (201) 539-7547. $ 2.50 postage and handling.)

US Environmental Protection Agency. *It's Not Over in October! A Guide for Local Emergency Planning Committees, Implementing the Emergency Planning and Community Right-to-Know Act (SARA Title III).* (Emergency Planning and Community Right-to-Know Information Hotline. Toll-free (800) 535 0202; in Washington, DC (202) 479-2449.

Accident Prevention

US Environmental Protection Agency Office of Solid Waste and Emergency Response (OS-120). *Why Accidents Occur: Insights From the Accidental Release Information Program and Chemical Accident Prevention Bulletin.* OSWER-89-008.1. July 1989. (Emergency Planning and Community Right-to-Know Information Hotline. Toll-free (800) 535-0202; in Washington, DC (202) 479-2449.

This publication provides an overview of accident and spill databases with a focus on the Accidental Release Information Program, its content, and the pilot program developed with ARIP data.

Environmental Action Foundation. *The Dynamic Duo: RCRA and SARA Title III: A Handbook on the Nation's Hazardous Waste Law* (Carol Dansereau). (Environmental Action Foundation, 6930 Carroll Avenue, Suite 600, Takoma Park, Maryland, 20912. (301) 891-1100. $10 for individuals and public interest groups, $20 for businesses or industry.)

Provides a comprehensive description of RCRA and how the law can be used in conjunction with the Emergency Planning and Community Right-to-Know Act.

US Environmental Protection Agency Office of Solid Waste (OS-312). *Hazardous Waste FOIA Reports.* (Available from this EPA office, OS-312, 401 M Street, SW, Washington, DC 20460. (202) 260-3393. Requests must be in writing or by fax: (202) 260-0295.)

The most comprehensive report to date describing publicly available RCRA information published. It describes the information available on the Resource Conservation and Recovery Information System (RCRIS) — a database that maintains information on the facilities regulated under RCRA.

US Environmental Protection Agency Office of Solid Waste (OS-312). *1991 Hazardous Waste Report Instructions and Forms.* EPA Form 8700-13A/B (5-80). Revised September 1991. (Available from this EPA office, OS-312, 401 M Street, SW, Washington, DC 20460.)

EPA prepared this reporting booklet for waste generators and treatment, storage, and disposal facilities to report their hazardous waste activities for 1991. You are advised to obtain a copy of this report if you are collecting waste minimization-related information on a plant.

Cooperation With Workers, Occupational Safety and Health

The following three National Toxics Campaign books contain useful information about cooperation with workers and obtaining occupational safety and health information. They are available from the National Toxics Campaign, 1168 Commonwealth Avenue, Boston, MA 02134. (617)232-0327.

National Toxics Campaign. *The Citizens Toxics Protection Manual.* ($35)

National Toxics Campaign. *Fighting Toxics, A Manual For Protecting Your Family, Community, and Workplace.* ($17)

National Toxics Campaign. The Good Neighbor Project of the Center for Public Policy. *The Good Neighbor Handbook: A Community-Based Strategy for Sustainable Industry* (Sanford Lewis). ($25, including shipping and handling, 15% discount for member groups.)

Mount Sinai School of Medicine of the City University of New York, Division of
Environmental and Occupational Medicine, Department of Community
Medicine. *Occupational Disease in New Jersey, Report to the New Jersey
Department of Health.* December 1989. (New Jersey Department of Health,
Right-to-Know Program, CN 368, Trenton, NJ 08625-0368. (609) 984-
2202. Free.)

*Chapter 3 contains a comprehensive analysis of occupational exposures to
hazardous materials in New Jersey.*

Occupational Safety and Health Act

The following OSHA publications and other occupational health and safety information are available from the US Department of Labor, Publications Office,
Room N3101, 200 Constitution Avenue, NW, Washington, DC 20210.

US Department of Labor Occupational Safety and Health Administration. *All
About OSHA.* 1991.

*Contains a step-by-step analysis of the law and its implementation in the Oc-
cupational Safety and Health Agency. It is particularly useful if your primary focus for the plant you are studying is on worker health and safety. It
will help you formulate your data request for both local and national OSHA
information.*

US Department of Labor Occupational Safety and Health Administration. *OSHA
Publications and Audiovisual Programs.* OSHA 2019. 1991.

*A complete guide to scores of booklets, documents, pamphlets, leaflets, and
audiovisual materials on occupational health and safety.*

Occupational Health Hazard Evaluations

US Department of Health and Human Services, Public Health Services, Centers
for Disease Control, National Institute for Occupational Safety and Health,
Division of Surveillance, Health Hazard Evaluations and Field Studies
(NIOSH). *The Health Hazard Evaluation Program of the National Institute
for Occupational Safety and Health.* (NIOSH, (800) 356-4674.)

*An overview of the process of requesting an HHE and blank copies of the
HHE itself. Although an HHE may be requested only by an employer or authorized employee representative, understanding the process can be a useful mechanism for learning about the outcome of the NIOSH investigation.
When ordering this document, also ask for NIOSH Health Hazard Evaluation Program (January 1992), a small but information-packed booklet that
provides a detailed description of the HHE Program.*

Chemical Manufacturers Association Responsible Care Program

When you call the CMA for Responsible Care information, ask for a list of member companies, a copy of the program elements, guiding principles, question and answer sheet, and list of general CMA resources (various books, pamphlets, and videos containing more information about CMA and the chemical industry). Although they may not yet be publicly available, ask CMA for copies of each of the Codes of Management Practices and sample copies of the member self-evaluation forms. You should also ask whether CMA has developed a policy manual on the issue of "product stewardship."

Chemical Manufacturers Association. *Preventing Pollution in the Chemical Industry, 1987-1990.* Spring 1992. (Chemical Manufacturers Association, 2501 M Street, NW, Washington, DC 20037. Free.)

US Environmental Protection Agency Office of Pollution Prevention and Toxics. *Pollution Prevention 1991, Progress on Reducing Industrial Pollutants.* EPA 21P-3003. October 1991. (Pollution Prevention Information Clearinghouse c/o SAIC 7600-A, Leesburg Pike, Room 369, Falls Church, VA 22043. (703) 821-4800; or fax your request to (703) 442-0584. Include document number, name, address, and telephone number.)

Chapter 3 contains specific information on four industrial trade association environmental programs, including the Chemical Manufacturers Association, American Petroleum Institute, National Paint and Coatings Association, and National Electrical Manufacturers Association.

Citizens Organizing Tools

The following two publications are available from the Working Group on Community Right-to-Know, 215 Pennsylvania Avenue, SE, Washington, DC 20003-1155. (202)546-9707. Make checks payable to RTK/PIRG Ed. Fund.

Center for Policy Alternatives, Working Group on Community Right-to-Know. *Making the Difference: Using the Right-to-Know in the Fight Against Toxics, Part I and Part II* (Nita Settina and Paul Orum). ($3 each for Part I [1990] and Part II [1991].)

Center for Policy Alternatives, Working Group on Community Right-to-Know. *Progress Report on Community Right-to-Know* (Paul Orum and Alair MacLean). 1992. ($6)

Citizens Clearinghouse for Hazardous Waste. *Research Guide for Leaders.* (Citizens Clearinghouse for Hazardous Waste, PO Box 926, Arlington, VA 22216. (703) 276-7070. $3.50.)

Explains how to research opponents, study corporations, analyze communities, and combine research with advocacy tactics.

Greenpeace and the Environmental Research Foundation (Ben Gordon and Peter Montague). *A Citizen's Toxic Waste Audit Manual.*(Greenpeace USA, 1436 U Street, NW, Washington DC 20009. (202) 462-1177. Free.)

How to obtain public information, particularly the Toxics Release Inventory database, for specific waste-generating facilities, and how to use the data in public campaigns for source reduction of toxic wastes.

The following three books are available from the National Toxics Campaign, 1168 Commonwealth Avenue, Boston, MA 02134. (617) 232-0327.

The Good Neighbor Project of the Center for Public Policy. *The Good Neighbor Handbook: A Community-Based Strategy for Sustainable Industry* (Sanford Lewis). ($25, including shipping and handling, 15% discount for member groups.)

Addresses issues surrounding the promotion of sustainability in industry; describes methods for developing and implementing "good neighbor agreements;" also contains case studies on successful citizen and labor campaigns.

National Toxics Campaign. *The Citizens Toxics Protection Manual.* ($35)

Chapters 4 and 5 describe researching and approaching a corporation. The manual also identifies data sources and the types of questions to ask about companies.

National Toxics Campaign. *Fighting Toxics, A Manual For Protecting Your Family, Community, and Workplace.* ($17)

The scope of this manual is also broader than source reduction, and it contains much information on corporate campaigns, neighborhood inspections, working with the media, federal environmental laws, and pollution prevention.

Technical Assistance by Environmental Professionals

The Environmental Careers Organization (ECO) is a national careers organization that places professionals in environmental fields. ECO is developing a new pollution prevention program known as the Technical Adviser Program for Toxics Use Reduction (TAPTUR). Technical advisers, such as retired scientists and engineers, assist nonprofit environmental and community groups by providing technical expertise and acting as mediators with industry and government. Such technical advisers can be a great help to pollution prevention efforts such as facility negotiations or the passage of pollution prevention laws.

For more information about ECO and TAPTUR contact:

Lori Colombo
Environmental Careers Organization
286 Congress Street
Boston, MA 02210-1009
(617) 426-4375

<table><tr><td>**Worksheet 1**</td><td>**Plant Identification and Profile**</td></tr></table>

Plant Background Information   Plant: _______________________________   Date filled out: _______________

(Attach copies of public documents to this worksheet or fill in the following information.)

A Identification

1 Plant name

2 Plant ownership or affiliation (include
   most recent changes in ownership)

3 Dun & Bradstreet number (this number
   may be useful as a plant identifier in
   public data searches)

B Output and Employment

1 Annual sales data (most recent year)

2 Annual production volume (most recent
   year)*

3 Number of plant employees in:
   (a) administration

   (b) production

4 Date operations began

5 Major changes in output and type of
   product in the past ten years

May not be available in public databases; this information may have to be obtained during the plant visit.   **continued...**

C Product Type

1 Categories of products manufactured
at the plant (include Standard Industrial Classification [SIC] code for product type; these codes are listed in
Table 5-1.)

2 Type of production process used: continuous or batch*

3 Changes planned in production output
that might proportionally affect the
quantity of toxic chemicals produced,
used, or released*

May not be available in public databases; this information may have to be obtained during the plant visit.* **end

Plant Background Information   Plant:________________________   Date filled out: ______________

A Permit Profile

1 Wastewater Discharge (specify pollutants)

Pollutant	Description of violation

2 Air Emissions (specify pollutants)

Pollutant	Description of violation

3 Hazardous Waste Treatment, Storage, Disposal Facility (TSD) Permit

Facility type	Capacity	Description of violation	Facility operating or closed

continued...

B Violations

1 Circumstances in which the plant
has exceeded limits established in
its permit

2 If the violation was sudden or accidental, what steps has the plant taken to
prevent a recurrence?

3 Have any permit violations led the plant
to take a closer look at source reduction?

C Permit Applications

1 Does the plant have any applications
pending before the state or local government for modifications to any existing permits? If so, what are they?

2 Does the plant have any applications
pending before the state or local government for a new permit (e.g., con-
struction of a new incinerator or land-
fill)? If so, what are they?

3 In reviewing the plant's application for
a permit modification or new permit,
has the state inquired whether a
source reduction program exists, or required one? If so, explain.

end

Plant Background Information   Plant: _________________________   Date filled out: _______________

(Attach a copy of the Form R data to this worksheet or fill in the following information.)

A Total Releases to the Environment On-Site and Off-Site

Chemical Name/CAS Number			
Air Fugitive emissions			
Stack emissions			
Total releases to air			
Water Stream or water body name			
Total releases to water			
Other on-site releases and disposal Underground injection			
Landfill			
Land treatment			
Surface impoundment			
Other disposal Off-site transfers			
Publicly owned treatment works (POTW)			
Total transferred to POTW			
Name and location of POTW			

Note: For each of the above areas, you may also want to ask company representatives about chemicals used, processed, or manufactured at the plant that are not on the TRI list or that are released below the TRI threshold.

continued...

A Total Releases to the Environment On-Site and Off-Site (continued)

Chemical Name/CAS Number			

Other transfers for treatment or disposal to off-site locations

Total transfers

Type of waste

Name and address of off-site location(s) for treatment, storage, or disposal

Other notes from the Form R or TRI data

B Recycling and Source Reduction

Chemical Name/CAS Number			

Plant Recycling, Energy Recovery, and Treatment Data

1 Quantity released

2 Quantity used for energy recovery
   on-site

3 Quantity used for energy recovery
   off-site

4 Quantity recycled on-site

5 Quantity recycled off-site

6 Quantity treated on-site

7 Quantity treated off-site

continued...

B Recycling and Source Reduction (continued)

Chemical Name/CAS Number			
Plant Recycling, Energy Recovery, and Treatment Data			
8 Quantity released into the environment as a result of remedial actions, catastrophic events, or one-time events not associated with production processes (pounds/year)			
9 Production ratio or activity index			
Source Reduction Information			
1 Calculation of quantity entering waste stream (For information on how to calculate this figure, see guidance for Worksheet 3, Section B: Recycling and Source Reduction, in Chapter 5)			
2 Reported source reduction activities: (a) activity (b) method			
3 Optional source reduction information reported			

continued...

C Commitments to US Environmental Protection Agency 33/50 Program

(Attach a copy of company's 33/50 commitment letter or fill in the following information.)

1 Has the company or the plant provided EPA with a written commitment (letter) to the 33/50 Program?

2 Does the letter describe the plant's voluntary reduction strategy? If so, does it include numerical reduction targets? What are they?

3 Which substances are targeted for reductions?

4 Have reductions been achieved? If so, how? (Look for source reduction methods.)

5 Does the company include reduction targets for other substances besides the seventeen 33/50 chemicals? What are those substances and corresponding reduction goals?

6 Notes on any other miscellaneous 33/50 information provided by the company to EPA

end

<table><tr><td>**Worksheet 4**</td><td>*Chemical Hazards and Chemical Storage*</td></tr></table>

Plant Background Information   Plant: _________________________   Date filled out: _________________

A Material Safety Data Sheets (MSDSs)
(Attach the MSDS(s) to this worksheet or fill in the following information for each chemical.)

1 Product identification

2 Warning statements

3 Precautionary measures

4 Emergency and first aid procedures

5 Occupational control procedures

6 Flammability and reactivity data

7 Health effects

8 Physical data

9 Spill, leak, and disposal information

B Chemical Inventories
(Attach copies of the inventories to this worksheet or fill in the following information for each chemical.)

1 Chemical description

2 Physical and health hazards

3 Inventory (maximum daily amount, average daily amount, number of days on
site)

4 Type of storage (i.e., container type)
and location at the facility

continued...

C New Chemical Substances (Premanufacture Notices) (OPTIONAL)

(Attach copies of the PMNs to this worksheet or fill in the following information for each new chemical substance.)

New Chemical Substance Name and CAS Number		
1 Production volume during the first 12 months of production		
2 Production volume for any consecutive 12-month period during the first three years of production		
3 Description of each intended category of use of the new chemical substance		
4 If "consumer" use is identified, include description of the new chemical substance in consumer products.		
5 If hazard information is identified, include hazard information reported or submitted (e.g., hazard warning statement, label, material safety data sheet, etc.).		
6 Name and address(es) of the industrial site(s) where the new chemical substance will be used		
7 Type of operation: batch or continuous?		

continued...

C New Chemical Substances (Premanufacture Notices) (OPTIONAL) (continued)
(Attach copies of the PMNs to this worksheet or fill in the following information for each new chemical substance.)

New Chemical Substance Name and CAS Number		
8 Occupational exposure to the new chemical substance:		
Worker activity		
Number of workers exposed		
Maximum duration: hours/day, days/year		
9 Environmental release and disposal:		
Amount of new chemical substances released		
Environmental medium (air, water, land)		
Control technology used to limit the release of the new chemical substance		
Water releases		
Name of POTW or navigable waterway		
NPDES number (permit)		
10 Optional pollution prevention information reported (describe in detail)		
11 Was the new chemical substance approved by the EPA? If not, why?		

end

Plant Background Information   Plant: _________________________________   Date filled out: ________________

A Community Emergency Preparedness Plan*
(Attach a copy of the plan to this worksheet or fill in the following information.)

1 What facilities are within the emergency planning district? What routes of transportation are likely to be used for the transport of hazardous substances? Are there other facilities (e.g., hospitals, schools, natural gas plants) that may contribute to or be subjected to additional risk due to their location near the subject facilities or routes of transportation of hazardous substances?

2 What are the methods and procedures to be followed by facility owners and operators and emergency and medical personnel to respond to releases of these substances?

3 Who are the community and emergency coordinators who will determine whether or not the plan will be implemented?

4 What are the procedures by which the facility emergency coordinators will provide notification to the public that a release has occurred?

* *Questions adapted from the federal Emergency Planning and Community Right-to-Know Act.*   **continued...**

A Community Emergency Preparedness Plan (continued)

5 What methods were used to determine the probability of a release and the area of the community that may be affected?

6 Describe the emergency equipment at each facility in the plan and identify the person(s) responsible for the equipment and facilities.

7 Is there an evacuation plan? What are the provisions for a precautionary evacuation and what (if applicable) are the alternative traffic routes?

8 Describe programs and schedules for training local emergency response and medical personnel.

9 What are the methods and schedules for carrying out the emergency plan?

continued...

Community Emergency Preparedness, Emergency Releases, and Accidents (continued)

B Accidental and Emergency Releases of "Extremely Hazardous Substances"*

Attach a copy of the incident report to this worksheet or fill in the following information for each substance.

Substance

1 Location of the release

2 Is the substance on the "extemely hazardous" list?

3 Quantity of the substance released

4 Time and duration of the incident

5 Medium of release:
Air
Water
Land
Combination

6 Known or anticipated health risks

7 Necessary medical attention

8 Proper precautions

9 Contact person at the facility

10 Update information for points covered in the initial notification

11 Description of response actions taken

12 More detailed information on health risks

13 Advice regarding medical care needed by exposure victims (if appropriate)

* *Questions and data elements adapted from the federal Emergency Planning and Community Right-to-Know Act.*

continued...

C Other Hazardous Substance Releases (Emergency Response Notification System [ERNS])

Substance		
1 Incident identification: Report number		
2 Material reported: Quantity spilled or released Media spilled or released into Waterway affected Description		
3 Incident location: Spill or release date State County City Location address		
4 Cause or source of incident:		
5 Damages/actions: Deaths Injuries Evacuations Damages Damage cost Action		

D Transportation Incidents (Hazardous Materials Information System [HMIS])

Attach a copy of the HMIS report to this worksheet or fill in the following information for each incident.

1 Date and time of incident		
2 Location of incident		
3 Mode of transportation (highway, rail, air, water, etc.)		
4 Description of carrier, company, or individual reporting (name, address, etc.)		

continued...

D Transportation Incidents (Hazardous Materials Information System [HMIS]) (continued)

5 Hazardous materials shipped

Hazardous materials involved in the incident

6 Consequences of incident due to the hazardous material:

Fatalities

Hospitalized injuries

Non-hospitalized injuries

Number of people evacuated

Estimated dollar amount of loss and/or property damage, including cost of decontamination or cleanup

7 Other description of event (additional information)

8 Name and telephone number of person preparing report

9 What precautions have been taken to prevent the recurrence of this or other incident involving hazardous materials?

10 Does the company have a specific process for evaluating carriers? Please describe.

continued...

E Accident Prevention (Accidental Release Information Program [ARIP])

[Note: Part of the information you receive in your Accidental Release Information Program (ARIP) search (facility profile and hazardous substance release profile) may contain some of the same information you obtain in your ERNS search. Thus, this section of the worksheet focuses only on the faciclity prevention profile.]

Fill in the following for each event.

1 What formalized evaluation was performed prior to this release at the process or storage area within your facility where the accident occurred? When was it last conducted? How frequently is this evaluation conducted?

2 Was the hazard evaluation performed effective in predicting this release event? Why or why not?

3 Identify the training procedures and/or management practices used at this facility prior to this release to prevent accidental releases. (Preventive maintenance/inspections? Accident investigations? Audits? Inventory/capacity reductions? Employee safety training? Standard operating procedures? Emergency response training? Other? Please describe.)

4 Describe any changes to existing training, procedures, and management practices, or any new types of training, procedures, and management practices, that are or will be implemented as a result of this release.

5 What engineering systems or controls were in use prior to the release at the process or storage area within the facility where the accident occurred?

6 Describe any changes to the existing engineering systems or controls, and any new types of engineering systems/ controls, that are or will be implemented as a result of the release.

7 Other notes from ARIP data

end

Plant Background Information   Plant: ________________________   Date filled out: ________________

(Attach copies of most recent Resource Conservation and Recovery Act (RCRA) Hazardous Waste Report biennial report data to this worksheet or fill in the following information.)

1 Name, title, and telephone number of person at the plant who should be contacted with questions about the hazardous waste report

2 Description of waste

3 Quantity of waste generated in 1990

   Quantity generated in 1991 (or most recent consecutive years reported)

4 Report any new activities that resulted in the minimization of this waste.

5 Did the facility begin or expand a source reduction activity? Recycling activity?

6 Did the facility investigate opportunities for source reduction or recycling?

7 What factors delayed or limited the site's ability to initiate new or additional source reduction activities?

8 What factors delayed or limited the facility's ability to initiate new or additional on-site or off-site recycling activities?

Note: You may also want to ask plant operators about waste streams that are not reported under RCRA.   **end**

Plant Background Information   Plant: _______________________   Date filled out: _______________

A Contacting Employee Representatives at the Plant*
(This information can be obtained through your contact with employee representatives.)

1 Contact person at plant.

2 Is the plant unionized? Number of employees covered.

3 What toxic or hazardous materials are used, produced, handled, or stored in the plant? Workers may know of chemicals used in the plant that are not reported to public databases.)

4 Are workers routinely or occasionally exposed to these materials? Please describe.

5 Are there any worker health and safety problems inside the plant that relate primarily to the use, production, handling, or storage of hazardous materials? Please describe. Have plant managers been notified of the problem(s)? What was the response?

6 Are there existing adverse impacts on the environment outside the plant? (Employees are often aware of environmental problems outside the plant that may not be publicly reported.) Please describe. Have plant managers been notified of the problem(s)? What was the response?

7 Do plant managers seek the input of employees in correcting health and safety in the plant? In improving environmental quality outside the plant? Are there specific institutional programs or policies that are designed to involve employees in this effort? Please describe.

* *Some of the suggested discussion issues for citizens groups and workers were adapted from* The Good Neighbor Handbook, A Community-Based Strategy for Sustainable Industry. *See Resources at the end of Chapter 5 under "Citizens Organizing Tools."*

continued...

A Contacting Employee Representatives at the Plant (continued)

8 Are employees involved in any pollution prevention/source reduction efforts in the plant? Please describe any employee incentive or award programs. Do employees take them seriously? Have any source reduction measures been investigated or implemented as a result of employee suggestions?

9 Is there a union safety and health committee?

10 Is there safety and health language in the collective bargaining agreement?

12 Are there regular plant inspections?

13 Do union officials want to be involved in or informed of citizen attempts to engage in dialogue with the plant regarding source reduction?

14 How can workers be helpful to environmental and community groups' efforts to promote source reduction?

15 How can community groups be helpful to worker involvement in source reduction?

continued...

B Occupational Safety and Health Act (OSHA) Data

1 In the past few years, has an OSHA inspection been conducted at the plant?
On what date(s)?

2 Was the inspection in response to an
owner or employee concern or com-
plaint? Describe. (Of primary interest
are complaints regarding existing or potential toxic hazards that could lead to
human exposure, such as leaks, spills,
explosions, etc.)

3 What was discovered in the inspection?

4 Was a union representative present at
the "walkaround" process inspection?

5 Was there any violation of law? Was a
citation issued? (Ask for a copy of the
citation(s) and the "narrative.") What
were the fines? When were the abate-
ment date(s)?

6 What was the final outcome? Was the
problem corrected? What precautionary measures have been taken or
planned to reduce the potential for
such a problem in the future?

7 Are there any summaries of OSHA problems at the plant which involve toxic or
hazardous substances (such as descriptions of exposures or other safety
issues involving hazardous materials)?

continued...

B Occupational Safety and Health Act (OSHA) Data (continued)

8 What is the current status of any other
   OSHA activity involving toxic or hazardous exposures at the plant?

C Occupational Health Hazard Information

(Attach a copy of the Health Hazard Evaluation (HHE) report or fill in the following information.)

1 Has an HHE which involves hazardous
   substance exposure or other chemical
   hazards been conducted at the plant?
   On what date(s)?

2 Why was an HHE requested? Describe
   the problem.

3 What hazardous substance(s) was involved?

4 How many workers were exposed or at
   risk?

5 What was the outcome of the investigation? (Was the problem corrected?
   Were preventive measures taken to reduce the potential for such a problem
   in the future?)

end

Plant Background Information   Plant: _________________________   Date filled out: _______________

(Contact state officials listed in Appendix B.)

1 Does the state have a source reduction
   program? (The program may have a different title, such as "pollution prevention," "toxics use reduction," "hazardous waste reduction," or "waste minimization." See Box 2-1 in Chapter 2 for
   a discussion of the different meanings
   of these terms.)

2 Was the program established pursuant
   to state law? If not, how was it established? How long has it been in operation?

3 Is there a source reduction or pollution
   prevention office which oversees the
   program?

4 Is there an independent advisory body
   for the program? (These are usually
   comprised of representatives of government, public interest or environmental
   groups, labor, the academic or science
   community, and industry.)

5 Who are the advisory board/council
   members, and how can they be contacted?

6 How is source reduction defined in the
   law or program?

7 What facilities are covered?

continued...

8 What chemicals are covered?

9 Does the law or program have mandatory or voluntary source reduction auditing and planning? Are the plans facility-wide or at the process level? Do they include reduction goals?

10 Does the facility report on annual progress? What information is publicly available? (Describe.) How can it be obtained?

11 Does this agency have written materials that describe the state source reduction program? How can they be obtained?

12 Other notes on your state source reduction program

end

<table><tr><td>Worksheet 9</td><td>Industrial Trade Associations – Chemical Manufacturers Association (CMA) Responsible Care Program</td></tr></table>

Plant Background Information   Plant: _______________________________   Date filled out: _______________

Questions for CMA

1 Is the facility a Chemical Manufacturers Association (CMA) member?

2 Name and telephone number of the facility contact

3 Can CMA send a copy of the Responsible Care "Guiding Principles, Program Elements" and copies of each of the six "Codes of Management Practices"? (Also ask for any other materials or publications they think would be useful to a citizens group.)

Questions for the facility contact

4 Has the company signed the "Guiding Principles" of the Responsible Care program?

5 Has the company had meetings or dialogue with local citizens or plant neighbors? Please describe. You may want to mention that you and your neighbors will be requesting a meeting to discuss source reduction at the plant.

continued...

Questions for the facility contact (continued)

6   Has the company provided plant tours to local citizens or neighbors? Please describe.

7   Has the company provided plant inspections to local citizens or neighbors? (An inspection is a plant tour that includes an expert of the local group's choosing who is trained to look for particular problems at the plant.)

8   Has the company completed an annual CMA self-evaluation form? Has it been submitted to CMA? May we have a copy? If no copy is available, would plant representatives be willing to review the issues in the self-evaluation forms with local citizens during the meeting with plant officials?

9   Other notes on the company's Responsible Care program

end

Plant Visit Information   Plant: _________________________________   Date filled out: _____________________

Company representative(s) _______________________________   Group representative(s) _____________________

A Establishing a Source Reduction Policy

1 Does your company have a working
   definition of source reduction? What is
   it? When was it first established?

2 Does your company have a written
   policy favoring source reduction as the
   primary strategy for protecting human
   health and the environment? If so,
   may we have a copy?

3 Who establishes the policy?

4 Is the policy corporate-wide, or plant-
   specific?

5 Which environmental media (air, water,
   and land) are addressed by the policy;
   that is, does it cover all gaseous, liquid, and solid substances used or generated at the plant?

6 What prompted the establishment of
   this policy?

7 Does the policy include sharing source
   reduction information with your suppliers, customers, or other industries?

continued...

A Establishing a Source Reduction Policy (continued)

8 Are source reduction options routinely considered during the engineering phase when planning new products or processes?

B Creating Company Leadership

1 Who is responsible for source reduction within your plant?

2 What is his/her/their title and technical/program management background? To whom do they report and who reports to them?

3 Is there a department, division, or team at the company or plant that is responsible for source reduction? If so, how is it staffed, where is it located, and what are its general responsibilities?

4 How does this entity fit into the overall corporate structure? To whom does it report and who reports to it?

C Involving Employees

1 How are employees involved in identifying opportunities, implementing changes, or measuring progress?

2 What is the role of managers?

continued...

C Involving Employees (continued)

3 What is the role of operators of your production processes?

4 Is there an incentive/reward system that encourages employees to come up with suggestions for source reduction? What are some examples of the incentives and rewards used (e.g., company awards, bonuses, posters)?

5 How are employees trained to recognize source reduction opportunities and put them to use? Is there a formal source reduction training program? If so, please describe.

6 How is source reduction progress transmitted throughout the plant/company (e.g., in the form of newsletters or staff meetings)? May we see examples of newsletters or other written materials?

7 Is employee involvement part of your written source reduction policy?

D Collecting Information

Materials Use and Handling:

1 Does the company collect throughput data; that is, does it track raw materials and products as they move sequentially from one end of the plant to the other? May we have a copy of the throughput data for the facility as a whole?

2 Are materials tracked in both production and non-production areas of the plant? (Nonproduction areas include storage, loading, transfer, and pollution control sites.)

continued...

D Collecting Information (continued)

Materials Use and Handling:

3 How are materials tracked? As specific chemicals?

   (circle all that apply and explain) By chemical categories?

   By waste categories?

   As products?

4 Which chemicals, categories, and/or wastes are tracked? Does the system track all types of materials (gases, liquids, and solids)?

5 Are all material types tracked regardless of whether they are subject to regulations when released to air, land, or water?

6 Does the tracking system include the activities of employees that might lead to waste generation (e.g., maintenance practices or the way raw materials are transferred from one area of the plant to another)? How?

7 Are statistical process controls (SPCs) used at this facility? If not, why not? If so, to which processes have they been applied?

Materials Balance Data: (optional)

1 Does the company go further by performing a detailed materials balance at each process? If so, what is the nature and level of uncertainty in the measurement methods used? How are these levels of uncertainty factored into the analysis of the overall results?

2 If no materials balance is performed, please explain why.

continued...

D Collecting Information (continued)

Gathering the Data:

1 How frequently are the data collected?

2 Who performs the materials tracking and/or materials balance procedures? What is his/her/their educational and professional background?

3 Have you computerized this information?

E Using the Collected Information

1 Who sees the results of the collected information (e.g., company leadership, the source reduction team, process managers)?

2 What are the data used for (e.g., to identify source reduction opportunities, improve process efficiency, track impact of employee training programs, identify health and safety concerns, ensure regulatory compliance)?

3 Have you established a base year from which present and future accomplishments are measured? If so, what is that year?

4 How are these baseline data modified to reflect changes in currently used processes, products, and production levels?

continued...

E Using the Collected Information (continued)

5 How do the collected data factor into
evaluations of both the source reduction
policy and the specific actions taken to
achieve reductions in waste generation?

F Assigning the Full Cost of Waste Back to the Source

1 Are waste-related costs assigned back to
their source in the company's accounting
system, or does the company treat environmental costs as a fixed overhead ex-
pense?

2 Which types of waste (air emissions,
wastewater discharges, and solid wastes)
are included in the cost accounting?

3 Are costs designated by: Specific chemicals?

(circle all that apply and explain) Chemical categories?

Waste categories?

4 What types of costs are captured by the
full cost accounting method you use?

a   Materials (e.g., the costs of
wasted starting materials and lost
products)?

b   Waste handling (e.g., capital and
operational expenses for on-site
recycling, treatment, storage, or
disposal facilities; transportation
and other expenses for wastes
sent off-site)?

c   Regulatory compliance?

d   Insurance?

e   Future liabilities from accidents,
worker illness, or waste site clean-
up?

f   Public/customer relations dealing
with waste issues?

g   Other (please specify)?

continued...

G Preparing Source Reduction Plans and Progress Reports

1 Does the plant prepare source reduction plans and progress reports?

2 Does your plan include specific source
reduction goals and timetables? If so,
what are the current goals and timetables?

3 Are source reduction plans based on: Specific chemicals?

(circle all that apply and explain)   Chemical categories?

Waste categories ?

4 Are they process-specific, plant-specific, or for the company as a whole?

5 How is source reduction progress measured?

By a decrease in pounds of chemical
used per pound of product produced?

By a decrease in pounds of chemical
waste generated per pound of product
produced?

By a decrease in pounds of chemical
waste generated per year for the facility
as a whole?

By a decrease in releases and transfers reported to EPA's Toxics Release
Inventory (TRI)?

Please use Worksheet 11 to report source reduction achievements and planned source reduction activities.   **end**

Plant Visit Information   Plant: _________________________   Date filled out: _____________

Plant representative ____

1 Source reduction activity (Describe)				
2 Date implemented				
3 Source reduction technique* (circle all that apply)	CH EQ OP PR PS		CH EQ OP PR PS	
4 Chemical(s) effected	Name	Name	Name	Name
Role and form of chemical in production process: G = gas, L = liquid, S = solid (circle all that apply)	Raw material G - L - S Product G - L - S Waste G - L - S	Raw material G - L - S Product G - L - S Waste G - L - S	Raw material G - L - S Product G - L - S Waste G - L - S	Raw material G - L - S Product G - L - S Waste G - L - S
5 Quantity of chemical constituent reduced: a. Pounds reduced annually as waste				
b. Pounds of waste reduced per unit of production				
c. Pounds reduced annually as raw material				
d. Pounds of raw material reduced per unit of production				
6 a. Percent reduction of waste generated				
b. Percent reduction of raw material used				
7 Capital costs				
8 Dollars saved (annually)				
9 Implementation time: Includes R&D? (circle one) If yes, R&D time	Yes No		Yes No	
10 Percent change in product yield (circle one)	% increase decrease		% increase decrease	
11 Motivating factor(s)				
12 How first identified?				

* *Key to source reduction techniques: CH, chemical substitution; EQ, equipment change; OP, operational change; PR, product change; PS, process change. See Chapter 3 for an explanation of these techniques.*   **end**

Plant Visit Information   Plant: **Fossil-to-Fabric**   Date filled out: **6/25/92**

Plant representative **Carmen Diockside**

1 Source reduction activity (Describe)	**Use purified intermediates so can reuse metallization filtrate in the process. Replaced a hazardous solvent w/ a non-hazardous substitute.**		**Install condensers on acrylonitrile-butadiene-styrene plastic production unit. Process changes enable additional recovery of monomer.**	
2 Date implemented	**1986**		**planned for next year**	
3 Source reduction technique*	(CH) EQ OP PR (PS)		CH (EQ) OP PR (PS)	
4 Chemical(s) affected, role and form of chemical in production process G = gas, L = liquid, S = solid (circle all that apply)	Name **heavy metals** Raw material G - L -(S) Product G - L - S Waste G -(L)(S)	Name **cellosolve acetate** Raw material G (L) S Product G - L - S Waste G (L) S	Name **acrylonitrile** Raw material G (L) S Product G - L - S Waste (G) L - S	Name Raw material G - L - S Product G - L - S Waste G - L - S
5 a. Pounds reduced annually as waste	**90,000**	**unknown**	**1 million**	
b. Pounds of waste reduced per unit of production	**18**	"	**less than 1**	
c. Pounds reduced annually as raw material	**0**	"	**1.5 million**	
d. Pounds of raw material reduced per unit of production	**18**	"	**less than 1**	
6 a. Percent reduction of waste generated	**80%**	**100%**	**50%**	
b. Percent reduction of raw material used	**0**	**100%**	**2%**	
7 Capital costs	**none**		**$60,000**	
8 Dollars saved (annually)	**$160,000**		**$1 million**	
9 Implementation time: Includes R&D? (circle one) If yes, R&D time	**6 months** (Yes) No		**less than 1 year** (Yes) No	
10 % change in product yield	**14** % (increase) decrease		**2** % (increase) decrease	
11 Production index	**14** % (increase) decrease		**0** % increase decrease	
12 Motivating factor(s)	**New wastewater discharge permit limits.**		**To comply with good neighbor agreement.**	
13 How first identified?	**Employee suggestion.**		**Plantwide program to identify fugitive emissions.**	

* *Key to source reduction techniques: CH, chemical substitution; EQ, equipment change; OP, operational change; PR, product change; PS, process change. See Chapter 3 for an explanation of these techniques.*   **end**

Your efforts to encourage local plants to adopt source reduction as their primary strategy for dealing with toxics problems can have a major impact on your community. If your interests are statewide, you will want to examine the role state and local officials are playing in promoting source reduction as government policy.

Government policies and activities can provide significant impetus to corporate progress in three ways: making source reduction a visible priority in the state's environmental protection strategy, getting companies to see more clearly that source reduction is in their own self-interest, and putting corporate practices under a strong public spotlight. Government involvement is vital to making source reduction the primary goal in national and state environmental programs and in shaping environmental policy. INFORM's research of 29 organic chemical plants in *Environmental Dividends: Cutting More Chemical Wastes* shows that environmental regulations were a prime motivating factor in seeking source reduction opportunities. The report showed a 300 percent increase in source reduction activities undertaken as a result of environmental regulations in the study plants between 1985 and 1990.

Even so, government regulators and industry leaders are still generally preoccupied with end-of-the-pipe solutions. In 1986, the congressional Office of Technology Assessment found that well over 99 percent of existing federal environmental resources were being spent on cleaning up pollution after it had been created. Although some progress has been made since then, pressing your state and federal representatives to promote source reduction aggressively is an important component of your effort to encourage source reduction.

In addition to the Pollution Prevention Act of 1990, which established source reduction as national policy, more than two dozen states have enacted source reduction, toxics use reduction, or pollution prevention laws since 1986. As of early 1993, such laws existed in Alaska, California, Connecticut, Delaware, Florida, Georgia, Illinois, Indiana, Iowa, Kentucky, Louisiana, Maine, Massachusetts, Minnesota, Mississippi, New Jersey, New York, North Carolina, Oregon, Rhode Island, Tennessee, Texas, Vermont, Washington, and Wisconsin. However, the scope and approach of the laws in each of these states vary substantially. The laws range from toxics use reduction (less chemical use) to waste minimization (reduction in the solid and semi-solid wastes regulated under the Resource Conservation and Recovery Act) to voluntary technical assistance programs.

More information about state laws may be obtained from the pollution prevention contacts in your state (listed in Appendix B). See also the publications listed in the Resources section at the end of this chapter. Worksheet 8 in Chapter 5 includes questions to ask government officials about source reduction programs in your state.

Some state officials may agree to meeting with community leaders to discuss the fine points of the state program. If they are not willing to meet, ask them to send written information which you can discuss with them on the telephone. If your state does not have a source reduction law, agency representatives should know whether voluntary programs promoting source reduction have been developed. The information in this chapter is designed to help you either evaluate an existing state program or promote its establishment.

In 1987, INFORM summarized six specific steps states can take to develop strong source reduction programs. These are not the only legislative approaches possible, but they represent one coherent approach that includes a combination of measures now widely considered in many states. This chapter is designed to give you a broad understanding of these six steps; INFORM's report, listed in the Resources, discusses them in more detail.

Six Steps to Source Reduction

1. Establish Source Reduction as the Top Priority

Source reduction should be established as explicit state policy, distinguished from waste management, pollution control, and recycling practices (see Chapter 2). Such a policy counteracts the strong tendency in industry and government to turn first to reactive pollution control solutions to toxic or hazardous waste problems.

As discussed in Chapter 2, even if source reduction is established as the first priority, and a facility does all it can to find source reduction opportunities, toxic or hazardous waste may still be created. These facilities must then turn to recycling or the most environmentally sound waste management method. Establishing this hierarchy as state policy sets the stage for the creation of a state source reduction program. If your state law already contains an explicit focus on source reduction or toxics use reduction, that law should be the springboard for your discussions with agency officials.

2. Establish a State Source Reduction Office

The state source reduction office should have three distinguishing characteristics if it is to operate effectively:

1. It should be highly placed in the government hierarchy, in, or closely associated with, the leading state environmental agency. This emphasizes the

priority the state places on the office's activities and promises top-level attention to policies designed to stimulate source reduction.

2. The office director should have a background in source reduction and industry production processes and command the respect of corporate leaders — preferably an environmentalist with corporate management experience.

3. The office must be separate from end-of-the-pipe regulatory offices that deal with individual environmental media. Such a limited setting would constrain its multimedia, preventive mandate and burden its agenda with the daily pressures of meeting regulatory timetables. Further, direct association with regulatory offices might interfere with the office's ability to create positive relationships with industry.

3. Delegate Nine Responsibilities to the Office

1. **Establish goals and timetables for source reduction** by identifying the industries and plants that should receive the highest-priority attention and by establishing general numerical goals and realistic timetables for source reduction within these industries. After several years of data analysis, the office should be able to provide the public with clear targets and priorities for source reduction.

2. **Review annual facility-level materials accounts, source reduction plans, and achievements.**

3. **Develop data collection programs** to assure that the state can establish program priorities and evaluate individual and overall progress toward source reduction. At a minimum, data should be collected on the amount and movement of specific toxic or hazardous chemicals. At the federal level, Title III Section 313 of the Superfund Amendments and Reauthorization Act and the Pollution Prevention Act of 1990 mandate that all plants meeting certain requirements report information on more than 300 toxic chemicals to the annual Toxics Release Inventory (see Appendix B). Some states have called for a broader list of chemicals. Massachusetts, for example, requires planning and reporting on about 1,000 chemicals in its toxics use reduction law.

   The key feature of a high-quality database is that it makes it possible to answer the central question: How has the amount of individual toxic pollutants released to the environment changed over a given year and how much of the change was due to source reduction as opposed to declines in production, increased treatment, or other strategies? Measurement of the actual source reduction achieved at a plant requires data on the amounts of stored chemicals at the beginning and end of the measurement period, the amounts entering the plant as raw materials, the amounts produced there, the amounts consumed in production, and the amounts leaving as product, as well as the amounts discharged as waste. New Jersey developed an efficient system for collecting this information in 1979. As a result, in 1986, New Jersey passed a state right-to-know law that includes publicly available facility-wide materials accounting data. Citizen groups in New Jersey used the information to convince their legislators and government agency officials of the need for mandatory source reduction planning and reporting for industrial facilities in the state. This effort led to the passage of the New Jersey Pollution Prevention Act in 1990. The New Jersey success stories illustrate how information collection can be a strong motivating factor in changing government policy.

It is also vitally important that the collected data be available to the public. Do not underestimate the importance of a sound public database for driving source reduction. In an area where standard-setting is very difficult, information is one of the most powerful tools for changing industry behavior.

Moreover, information on heavy waste loads publicly identifies operators of polluting facilities. Companies spend millions of dollars annually on advertising to strengthen their public image. They do not want that image threatened by negative publicity. On the other hand, public acknowledgment of those companies that recognize the importance of sound environmental practices and go beyond regulatory requirements with their source reduction efforts is important. It rewards the "good actor" and can spur competition between companies for future public recognition.

Most important, a good publicly available database will provide you and your state government with a comprehensive understanding of pollution problems and pollution controls in your state. This information has been sorely lacking at the state and federal levels; its absence makes the creation of sound policy decisions difficult, at best.

4. **Identify regulatory and nonregulatory impediments** to source reduction. Often, regulations and loopholes can create inexpensive pollution management alternatives that are dangerous to the environment or public health and that serve to discourage source reduction (deep well injection, or disposal of waste deep underground, is one example). State funding priorities, mechanisms, and sources, as well as state tax policies, can also encourage or discourage source reduction. The state source reduction office should coordinate activities with the "single medium" offices of the state environmental agency to make sure that source reduction options are considered first.

5. **Analyze findings of source reduction research,** evaluate the need for research on source reduction options within particular industrial processes or categories of chemical pollutants, and promote research programs to meet these needs.

6. **Sponsor meetings** with corporate, trade association, union, environmental, and other leaders to promote source reduction. State officials could use these meetings to promote the company source reduction practices discussed in Chapter 3.

7. **Establish a program of on-site technical assistance** within the government, universities, trade associations, or other organizations to help smaller companies and isolated industries identify source reduction opportunities. This program could provide assistance with source reduction inventories if requested. It could also develop a computerized database in cooperation with other states to share information on specific source reduction technologies and approaches.

8. **Issue regular reports** on the state's progress in reducing pollution, priority industries for source reduction activities, and the office's activities, achievements, and future goals.

9. **Actively pursue federal funding** for source reduction activities. The US Environmental Protection Agency's Office of Pollution Prevention periodically makes funds available for state technical assistance programs (see Resources at the end of the chapter). Make sure your state program takes advantage of any available funds to build a strong infrastructure that promotes multimedia source reduction, and not just compliance strategies for Resource Conservation and Recovery Act (RCRA) regulations.

4. Establish "Right-to-Know" Policies

State "right-to-know" requirements, and "right-to-know" officers in the state regulatory agency, can be important in ensuring that industrial plant managers provide the information needed by public groups to evaluate corporate source reduction strategies. Encourage your state to pass a "sunshine" law, or your governor to issue a sunshine proclamation, directing industry to cooperate with the public's efforts to bring source reduction information out into the open. This could be done in several ways:

1. State policy should require industrial plants to make public the data needed to measure source reduction progress and target source reduction opportunities. For instance, companies should provide information on fluctuations in production levels, amounts of all pollutants leaving the facility, generic types of source reduction used at the company, reductions in toxics use and by-product generation, and amounts of waste no longer created. While trade secrets would still be withheld, state policy should condemn irresponsible use of that privilege.

2. Establish statewide awards for companies that cooperate.

3. Designate a right-to-know officer within the state regulatory agency to help citizens obtain data on individual companies, including Toxics Release Inventory information and data on the status of pollution permits. This officer could also tally information from citizens groups on companies that refuse to work with them at all.

5. Ensure That All Regulatory Programs Promote Source Reduction

There are a variety of opportunities at both the state and federal level for a state source reduction office to ensure a source reduction focus in existing regulatory programs.

1. The office should encourage personnel in single-medium regulatory programs to treat source reduction as the primary environmental protection strategy. For example, it should push for aggressive implementation of the federal 1984 Resource Conservation and Recovery Act Amendments that require waste generators to certify that they have a reduction program and to report biennially on their efforts to reduce the volume and/or toxicity of their wastes and the results they have achieved.

2. The office can provide source reduction training and materials to state and local regulatory personnel responsible for permitting and enforcement. It may be impractical to develop an independent framework for regulatory source reduction standards, but many opportunities exist to weave source reduction options into the fabric of pollution control permits required under existing programs. Teaching plant inspectors to carry out multimedia inspections, instead of single-medium inspections as they do today, is another key to ensuring that current laws are used to promote source reduction.

3. The office should discourage cheap and more environmentally risky waste disposal activities such as deep well injection of wastes and combustion of toxic wastes.

4. The office can work to expand the number of chemical substances that are regulated. Approximately 400 wastes or waste streams are regulated under RCRA. The Clean Air Act regulates only 189 toxic substances and six so-called "criteria pollutants" (sulfur dioxide, carbon monoxide, particulates, nitrogen oxides, ozone, and lead). Under the Clean Water Act, just 126 substances are regulated as so-called "priority pollutants" (toxic chemicals that are particularly harmful to one or more forms of animal or plant life, grouped into organics and metals). These are nearly insignificant numbers in the face of the 70,000 chemicals used commercially today.

5. Finally, the office can urge stricter enforcement of all environmental laws. INFORM's research of source reduction activities at 29 organic chemical plants revealed that even environmental regulations that did not directly require source reduction action, such as stricter limitations on transfers of hazardous or toxic waste to publicly owned treatment works, were a significant factor in motivating source reduction activity. In fact, environmental regulations were the fastest-growing source reduction incentive at these plants in recent years, cited as the reason for 40 source reduction activities implemented between 1985 and 1990 (a period when more environmental regulations were being enforced), as opposed to 10 activities before 1985 — a 300 percent increase.

6. Establish State Programs Supporting Source Reduction Policies

Beyond encouraging industry to adopt source reduction strategies, the state can consider instituting source reduction "sticks" to complement voluntary "carrots." Several methods are available.

1. The state could require any facility requesting permit approval under the Resource Conservation and Recovery Act, Clean Water Act, Clean Air Act, or Toxic Substances Control Act to submit source reduction inventories and plans.

2. The state could require that efforts to meet permitted discharge levels exploit source reduction options before using pollution management techniques.

3. The state could consider imposing a comprehensive pollution fee. In order to promote source reduction, the fee must be applied to the quantity of waste generated prior to any handling or treatment, regardless of the manner in which it is handled or the environmental medium into which it is discharged. Pollution must be quantified by the weight of the specific chemicals discharged, not by the volume of waste in which the chemicals are contained (e.g., activities such as dewatering which reduce the waste volume but not the chemical constituents would not qualify for a reduced fee).

    Such a fee leaves no loopholes, because it covers discharges to all media. It is equitable, because industries producing the largest amount of toxic wastes pay the most money. And it avoids the problems associated with narrower end-of-the-pipe taxes proposed to date. Using RCRA hazardous waste data as a basis for fees, for example, encourages companies to shift wastes into routes of discharge that are not regulated under RCRA, rather than reduce the amount of wastes. This can also make state revenue projections unreliable.

    A comprehensive pollution fee structure could involve a number of variations. The fee could vary based upon the level of toxicity of each pollutant, or discharges directly to the environment could involve higher fees than discharges to treatment facilities.

4. The state could consider phasing out (**sunsetting**) or banning particular chemicals. Although phase-outs and bans of chemicals can be controversial, because of the economic and labor issues involved, the concepts are gradually receiving wider acceptability. For example, a process to phase out or "sunset" hazardous chemicals has been discussed for the Great Lakes region. Because of the widespread environmental contamination caused by more than 500 hazardous chemicals used and discharged in the Great Lakes basin, the US/Canadian International Joint Commission recommended in its sixth annual report the use of a sunsetting process to achieve virtual elimination of hazardous chemicals.[1] Although much work is still to be done in this area, your state officials should be aware of the current status of the issue and, if possible, consider whether state resources or research might contribute to the progress of the sunsetting debate.

Before you advocate fees, permitting requirements, or a sunsetting effort, consider enlisting your state governor in a cooperative strategy. Organize a meeting of the governor, key industrial representatives, environmentalists, labor representatives, and technical experts in which the governor identifies goals for reducing pollutants in three to five of the state's most waste-intensive industries. The governor can make a direct, public challenge to industries — possibly with extensive press coverage — to either meet reduction goals in a given time period through their own private strategies, or face fees and/or permit requirements imposed by the state. This strategy allows you to test the spirit of cooperation, and it gives industries the opportunity to demonstrate that they can take the initiative. It also may create a political climate conducive to stricter requirements if industries fail to meet the challenge.

Resources

The following two publications are available from INFORM. See p. 198 for ordering and discount information.

INFORM (Warren R. Muir and Joanna Underwood). *Promoting Hazardous Waste Reduction: Six Steps States Can Take.* 1987. ($3.50)
This brief guide on state programs was the basis for this chapter.

INFORM (Mark H. Dorfman, Warren R. Muir, PhD., Catherine G. Miller, PhD.). *Environmental Dividends: Cutting More Chemical Wastes.* 1992. ($75)
This study of the same plants covered in Cutting Chemical Wastes offers a comprehensive analysis of 181 source reduction activities at these organic chemical plants. The analysis includes detailed descriptions of each source reduction activity, quantities of waste reduced, changes in production yield, economic savings, motivation for initiating source reduction activity, and implementation time. Much of the supporting data cited in this chapter was from this study.

Local Government Commission, Inc. (Anthony Eulo, Toxics Policy Director). (Suite 205, 909 12th Street, Sacramento, CA 95814. (916) 448-1198.)
Although not discussed in this chapter, promoting source reduction through local government programs has been successful in California, where county governments are particularly strong. Anthony Eulo can give you information on local government efforts and how to enhance them.

National Toxics Campaign. *Policy and Program Options for Reduction of Hazardous Waste in Texas.* (National Toxics Campaign, 1168 Commonwealth Avenue, Boston, MA 02134. (617) 232-0327. $17.)
Although this document was prepared specifically for Texas, it contains information of general interest. The section called "Six Aspects of a Compre-

hensive Waste Reduction Program" was this guide's source for information
on integrating inspections and phasing out or banning chemicals.

Office of Technology Assessment. *Serious Reduction of Hazardous Waste.* (Summary available from the Office of Technology Assessment, United States Congress, Washington, DC 20510. (202) 224-8996. Free. Book available from the National Technical Information Service, 5285 Port Royal Road, Springfield, VA 22161. $28.95 in paperback or $6.95 on microfiche, plus $3 per order handling for either edition.)

US Environmental Protection Agency. Office of Pollution Prevention and Toxics. (EPA Headquarters, (202) 260-6238.)

This office has information on federal funds available for state technical assistance activities.

Ventura County Environmental Health Program. *Hazardous Waste Reduction Guidelines for Environmental Health Programs.* May 1987. (Ventura County Environmental Health Program, 800 S. Victoria Ave., Ventura, CA 93009. (805) 654-2813. $5 at counter, $9 via mail.)

A guide to all aspects of local government programs.

Waste Reduction Institute for Training and Applications Research (WRITAR). *Survey and Summary of State Legislation Relating to Pollution Prevention.* 1992. (Available from WRITAR, 1313 5th Street, SE, Suite 325, Minneapolis, MN 55414-4502. $25 plus $2.50 shipping and handling.)

An analysis of legislation in 27 states that is designed to promote pollution prevention and waste management.

Notes

1. International Joint Commission, Sixth Biennial Report on Great Lakes Water Quality, Ottawa, Canada, and Washington, DC, April 1992.

Appendix A State Contacts for TRI Information

Alabama

E. John Williford, Chief of Operations
Alabama Emergency Response Commission
Alabama Department of Environmental Management
1751 Congressman W.L. Dickinson Drive
Montgomery, AL 36109
(205) 271-7931

Alaska

Amy Skilbred
Alaska State Emergency Response Commission
P.O. Box O
Juneau, AK 99811-1800
(907) 465-2630

American Samoa

Pati Faiai, Director
American Samoa EPA
Office of the Governor
Pago Pago, AS 96799
International Number (684) 633-2304

Arizona

Carl F. Funk, Executive Director
Arizona Emergency Response Commission
Division of Emergency Services
5636 East McDowell Road
Phoenix, AZ 85008
(602) 231-6326

Arkansas

Anna Brannon
Depository of Documents
Arkansas Department of Labor
10421 West Markham
Little Rock, AR 72205
(501) 682-4541

California

Chuck Shulock
Office of Environmental Affairs
P.O. Box 2815
Sacramento, CA 95812
Attn: Section 313 Reports
(916) 324-8124
(916) 322-7236 Completed Form R Information

Colorado

Colorado Emergency Planning Commission
Colorado Department of Health
4210 East 11th Avenue
Denver, CO 80220
Attn: Judy Waddill
(303) 331-4858

Commonwealth of Northern Mariana Islands

Frank Russell Meecham, III
Division of Environmental Quality
P.O. Box 1304
Saipan, CNMI 96950
(670) 234-6984

Connecticut

Sue Vaughn, Title III Coordinator
State Emergency Response Commission
Department of Environmental Protection
State Office Building, Room 161
165 Capitol Avenue
Hartford, CT 06106
(203) 566-4856

Delaware

Robert French, Chief Program Administrator, Air
Resource Section
Department of Natural Resources and Environmental
Control
89 King's Highway
P.O. Box 1401
Dover, DE 19903
(302) 739-4791

District of Columbia

Frank Jasmine,
District of Columbia Emergency Response
Commission
Office of Emergency Preparedness
2000 14th Street, NW
Frank Reeves Center for Municipal Affairs
Washington, DC 20009
(202) 727-6161

Florida

Jim Loomis
Florida Emergency Response Commission
Florida Department of Community Affairs
2740 Centerview Drive
Tallahassee, FL 32399-2149
(904) 488-1472
In Florida: (800) 635-7179

Georgia

Jimmy Kirkland
Georgia Emergency Response Commission
205 Butler Street, SE
Floyd Tower East
11th Floor, Suite 1166
Atlanta, GA 30334
(404) 656-6905

Guam

Roland Solidio
Guam EPA
P.O. Box 2999
Aguana, GU 96910
(671) 646-8863

Hawaii

John C. Lewin, MD, Chairman
Hawaii State Emergency Response Commission
Hawaii State Department of Health
P.O. Box 3378
Honolulu, HI 96801-9904
(808) 548-6505

Idaho

Idaho Emergency Response Commission
State House
Boise, ID 83720
(208) 334-5888

Illinois

Carlita Crockett
Emergency Planning Unit
Illinois EPA
Office of Chemical Safety #29
P.O. Box 19276
Springfield, IL 62794-9276
(217) 524-1006

Indiana

Philip Powers, Director
Indiana Emergency Response Commission
5500 West Bradbury Avenue
Indianapolis, IN 46241
(317) 243-5176

Iowa

Department of Natural Resources
Records Department
900 East Grand Avenue
Des Moines, IA 50319
(515) 281-8852

Kansas

Right-to-Know Program
Kansas Department of Health and Environment
Mills Building, 5th Floor
109 SW 9th Street
Topeka, KS 66612
(913) 296-1690

Kentucky

Valerie Hudson
Kentucky Department of Environmental Protection
18 Reilly Road
Frankfort, KY 40601
(502) 564-2150

Louisiana

R. Bruce Hammatt
Emergency Response Coordinator
Department of Environmental Quality
P.O. Box 44066
333 Laurel Street
Baton Rouge, LA 70804-4066
(504) 342-8617

Maine

Dorean Maines
State Emergency Response Commission
State House Station Number 11
157 Capitol Street
Augusta, ME 04333
(207) 289-4080
In Maine: (800) 452-8735

Maryland

Marsha Ways
State Emergency Response Commission
Maryland Department of the Environment
Toxics Information Center
2500 Broening Highway
Baltimore, MD 21224
(301) 631-3800

Massachusetts

Arnold Sapenter
c/o Title III Emergency Response Commission
Department of Environmental Quality Engineering
1 Winter Street, 10th Floor
Boston, MA 02108
(617) 292-5993

Michigan

Title III Coordinator
Michigan Department of Natural Resources
Environmental Response Division
Title III Notification
P.O. Box 30028
Lansing, MI 48909
(517) 373-8481

Minnesota

Lee Tishler, Director
Minnesota Emergency Response Commission
190 Bigelow Building
450 North Syndicate
St. Paul, MN 55104
(612) 643-3000

Mississippi

J.E. Maher, Chairman
Mississippi Emergency Response Commission
Mississippi Emergency Management Agency
P.O. Box 4501
Fondren Station
Jackson, MS 39296-4501
(601) 960-9973

Missouri

Dean Martin, Coordinator
Missouri Emergency Response Commission
Missouri Department of Natural Resources
P.O. Box 3133
Jefferson City, MO 65102
(314) 751-7929

Montana

Tom Ellerhoff, Co-Chairman
Montana Emergency Response Commission
Environmental Sciences Division
Department of Health & Environmental Sciences
Cogswell Building A-107
Helena, MT 59620
(406) 444-6911

Nebraska

Clark Smith, Coordinator
Nebraska Emergency Response Commission
Nebraska Department of Environmental Control
P.O. Box 98922
State House Station
Lincoln, NE 68509-8922
(402) 471-2186

Nevada

Bob King, Director
Division of Emergency Management
2525 South Carson Street
Carson City, NV 89710
(702) 885-4240

New Hampshire

George L. Iverson, Director
State Emergency Management Agency
Title III Program
State Office Park South
107 Pleasant Street
Concord, NH 03301
(603) 271-2231

New Jersey

New Jersey Emergency Response Commission
SARA Title III Section 313
Department of Environmental Protection
Division of Environmental Quality
Bureau of Hazardous Substances Information
CN-405
Trenton, NJ 08625
(609) 292-6714

New Mexico
Samuel Larcombe
New Mexico Emergency Response Commission
New Mexico Department of Public Safety
P.O. Box 1628
Santa Fe, NM 87504-1628
(505) 827-9222

New York
New York Emergency Response Commission
New York State Department of Environmental
    Conservation
Bureau of Spill Response
50 Wolf Road, Room 326
Albany, NY 12233-3510
(518) 457-4107

North Carolina
North Carolina Emergency Response Commission
North Carolina Division of Emergency Management
116 West Jones Street
Raleigh, NC 27603-1335
(919) 733-3867

North Dakota
SARA Title III Coordinator
North Dakota State Department of Health and
    Consolidated Laboratories
1200 Missouri Avenue
P.O. Box 5520
Bismarck, ND 58502-5520
(701) 224-2374

Ohio
Cindy Sferra-DeWulf
Division of Air Pollution Control
1800 Watermark Drive
Columbus, OH 43215
(614) 644-2266

Oklahoma
Larry Gales
Oklahoma Department of Health
Environmental Health Services Division
P.O. Box 53551
Oklahoma City, OK 73152
(405) 271-8056

Oregon
Ralph M. Rodia
Oregon Emergency Response Commission
c/o State Fire Marshall
3000 Market Street Plaza, Suite 534
Salem, OR 97310
(503) 378-2885

Pennsylvania
James Tinney
Bureau of Right-to-Know
Room 1503
Labor and Industry Building
7th & Forrester Streets
Harrisburg, PA 17120
(701) 783-2071

Puerto Rico
SERC Commissioner
Title III-SARA Section 313
Puerto Rico Environmental Quality Board
P.O. Box 11488
Santurce, PR 00910
(809) 722-0077

Rhode Island
Department of Environmental Management
Division of Air and Hazardous Materials
291 Promenade Street
Providence, RI 02908
Attn: Toxics Release Inventory
(401) 277-2808

South Carolina
Ron Kinney
Department of Health and Environmental Control
2600 Bull Street
Columbia, SC 29201
(803) 734-5200

South Dakota
Lee Ann Smith, Director
South Dakota Emergency Response Commission
Department of Water and Natural Resources
Joe Foss Building
523 East Capitol
Pierre, SD 57501-3181
(605) 773-3153

Tennessee

Lacy Suiter, Chairman
Tennessee Emergency Response Commission
Director, Tennessee Emergency Management Agency
3041 Sidco Drive
Nashville, TN 37204
(615) 252-3300
(800) 262-3300 (in Tennessee)
(800) 258-3300 (out of state)

Texas

David Barker, Supervisor
Emergency Response Unit
Texas Water Commission
P.O. Box 13087-Capitol Station
Austin, TX 78711-3087
(512) 463-8527

Utah

Neil Taylor
Utah Hazardous Chemical Emergency Response
  Commission
Utah Division of Environmental Health
288 North 1460 West
P.O. Box 16690
Salt Lake City, UT 84116-0690
(801) 538-6121

Vermont

Dr. Jan Carney, Commissioner
Department of Health
60 Main Street
P.O. Box 70
Burlington, VT 05402
(802) 863-7281

Virginia

Harry E. Gregori, Jr.
Virginia Emergency Response Council
Department of Waste Management
James Monroe Building, 14th Floor
101 North 14th Street
Richmond, VA 23219
(804) 225-2997

Virgin Islands

Allan D. Smith, Commissioner
Department of Planning and Natural Resources
U.S. Virgin Islands Emergency Response
  Commission
Title III
Nisky Center, Suite 231
Charlotte Amalie
St. Thomas, VI 00802
(809) 774-3320, Ext. 169 or 170

Washington

Chuck Clark, Chairman
Department of Community Development
9th and Columbia Building
Mail Stop GH-51
Olympia, WA 98504
(206) 753-2200

West Virginia

Carl L. Bradford, Director
West Virginia Emergency Response Commission
West Virginia Office of Emergency Services
State Capital Building 1, Room EB-80
Charleston, WV 25305
(304) 348-5380

Wisconsin

Department of Natural Resources
P.O. Box 7921
Madison, WI 53707
Attn: Russ Dumst
(608) 266-9255

Wyoming

Ed Usui, Executive Secretary
Wyoming Emergency Response Commission
Wyoming Emergency Management Agency
Comprehensive Emergency Management
P.O. Box 1709
Cheyenne, WY 82003
(307) 777-7566

Source: These names and addresses are provided by the
Environmental Protection Agency in *Toxics in the Community:
National and Local Perspectives,* EPA 560/4-91-014,
September 1991.

B State Contacts for Pollution Prevention Information

Alabama

Alabama Waste Reduction and Technology Transfer (Wratt) Program
Daniel E. Cooper, Chief
Special Projects
Alabama Department of Environmental Management
1741 Congressman William L. Dickinson Drive
Montgomery, AL 36130
(205) 271-7939

Alaska

Pollution Prevention Office
David Wigglesworth, Chief
Pollution Prevention Office
Alaska Department of Environmental Conservation
P.O. Box O
Juneau, AK 99811-1800
(907) 465-5275

Arizona

Arizona Waste Minimization Program
Stephanie Wilson
Dr. J. Andy Soesilo
Arizona Waste Minimization Program
Arizona Department of Environmental Quality
2005 North Central Avenue
Phoenix, AZ 85004
(602) 257-2318/6995

Arkansas

Arkansas Pollution Prevention Program
Robert J. Finn
Hazardous Waste Division
Arkansas Department of Pollution Prevention and Ecology
P.O. Box 8913
Little Rock, AR 72219-8913
(501) 570-2861

California

Department of Toxic Substances Control
Kim Wilhelm
Department of Toxic Substances
Alternative Technology Division
400 P Street
P.O. Box 806
Sacramento, CA 95812-0806
(916) 324-1807

Department of Toxic Substances Control
Tony Eulo
Local Government Commission
909 12th Street
Suite 205
Sacramento, CA 95814
(916) 448-1198

Colorado

Pollution Prevention and Waste Reduction Program
Neil Kolwey
Colorado Department of Health
4210 East 11th Avenue
Denver, CO 80220
(303) 331-4830

Connecticut

Connecticut Department of Environmental Protection
Carmine Di Battista, Director
Planning and Standards
Waste Management Bureau
Connecticut Department of Environmental Protection
165 Capitol Avenue
Hartford, CT 06106
(203) 566-3437

Connecticut Department of Environmental Protection
Elizabeth Flores, Assistant Director
Waste Engineering and Enforcement Division
Waste Management Bureau
Connecticut Department of Environmental Protection
165 Capitol Avenue
Hartford, CT 06106
(203) 566-8843

Delaware

Delaware Pollution Prevention Program
Philip J. Cherry
Andrea K. Farrell
Pollution Prevention Program
Department of Natural Resources and Environmental
    Control
P.O. Box 1401
Kings Highway
Dover, DE 19903
(302) 739-5071/3822

District of Columbia

Office of Recycling
Hampton Cross, Acting Recycling Coordinator
Office of Recycling
District of Columbia Department of Public Works
65 K Street, Lower Level
Washington, DC 20002
(202) 939-7116

Florida

Waste Reduction Assistance Program (WRAP)
Janeth A. Campbell
Waste Reduction Assistance Program
Florida Department of Environmental Regulation
2600 Blair Stone Road
Tallahassee, FL 32399-2400
(904) 488-0300

Georgia

Georgia Multimedia Source Reduction and Recycling
    Program
Susan Hendricks, Environmental Specialist
Environmental Protection Division
Georgia Department of Natural Resources
Floyd Tower East, Suite 1154
205 Butler Street, SE
Atlanta, GA 30334
(404) 656-2833

Hawaii

Hazardous Waste Minimization Program
Jane Dewell
Waste Minimization Coordinator
State of Hawaii Department of Health
Solid and Hazardous Waste Branch
5 Waterfront Plaza, Suite 250
500 Ala Moana Boulevard
Honolulu, HI 96813
(808) 586-4226

Idaho

Division of Environmental Quality
Joy Palmer
Katie Sewell
Division of Environmental Quality
Idaho Department of Health and Welfare
1410 North Hilton Street
Boise, ID 83720-9000
(208) 334-5879

Illinois

Office of Pollution Prevention
Mike Hayes
Illinois Environmental Protection Agency
Office of Pollution Prevention
2200 Churchill Road
P.O. Box 19276
Springfield, IL 62794-9276
(217) 785-0533

Indiana

Office of Pollution Prevention and Technical
    Assistance
Joanne Joice, Director
Charles Sullivan, Environmental Manager
Office of Pollution Prevention and Technical
    Assistance
Indiana Department of Environmental Management
105 South Meridian Street
P.O. Box 6015
Indianapolis, IN 46225
(317) 232-8172

Iowa

Waste Management Authority Division
Tom Blewett, Environmental Specialist
Waste Management Authority Division
Department of Natural Resources
Wallace State Office Building
Des Moines, IA 50319
(515) 281-8941

Kansas

State Technical Action Plan(STAP)
Tom Gross, Bureau Chief
State Technical Action Plan
Kansas Department of Health and Environment
Forbes Field, Building 740
Topeka, KS 66620
(913) 296-1603

Kansas State University Ritta Program
Lani Himegarner, Program Manger
Engineering Extension Programs
133 Ward Hall
Kansas State University
Manhattan, KS 66506-2508
(913) 532-6026

Kentucky

Kentucky Partners – State Waste Reduction Center
Joyce St. Clair, Executive Director
Kentucky Partners – State Waste Reduction Center
Ernst Hall, Room 321
University of Louisville
Louisville, KY 40292
(502) 588-7260

Louisiana

Louisiana Department of Environmental Quality
Gary Johnson, Waste Minimization Coordinator
P.O. Box 82263
Baton, Rouge, LA 70884-2263
(504) 765-0720

Maine

Bureau of Oil and Hazardous Materials Control
Scott Whittier, Director
Bureau of Oil and Hazardous Materials Control
Department of Environmental Protection
State House Station #17
Augusta, ME 04333
(207) 289-2651

Maryland

Office of Waste Minimization and Recycling
Harry Benson, Chief
Office of Waste Minimization and Recycling
Hazardous and Solid Waste Management
    Administration
Maryland Department of the Environment
2500 Broening Highway, Building 40
Baltimore, MD 21224
(301) 631-3315

Massachusetts

Office of Technical Assistance For Toxics Use
    Reduction
Barbara Kelley, Director
Richard Reibstein, Outreach Director
Massachusetts Department of Environmental
    Management
100 Cambridge Street
Boston, MA 02202
(617) 727-3260

Toxics Use Reduction Institute
Jack Luskin, Director of Education and Outreach
Toxics Use Reduction Institute
University of Lowell
1 University Avenue
Lowell, MA 01854
(508) 934-3275

Michigan

Office of Waste Reduction Services
Larry E. Hartwig, Director
Office of Waste Reduction Services
Michigan Department of Commerce and Natural
    Resources
116 West Allegan Street
P.O. Box 30004
Lansing, MI 48909
(517) 335-1178

Minnesota

Minnesota Pollution Control Agency (MPCA)
Eric Kilberg, Pollution Prevention Coordinator
Environmental Assessment Office
Minnesota Pollution Control Agency
520 Lafayette Road
St. Paul, MN 55155
(612) 296-8643

Mississippi

Mississippi Technical Assistance Program
(MISSTAP) and Mississippi Solid Waste
Reduction Assistance Program (MISSWRAP)
Dr. Caroline Hill
Mississippi Technical Assistance Program and
Mississippi Solid Waste Reduction Assistance
P.O. Drawer CN
Mississippi State, MS 39762
(601) 325-8454

Mississippi Technical Assistance Progran
(MISSTAP) and Mississippi Solid Waste
Reduction Assistance Program (MISSWRAP)
Thomas E. Whitten, Director
Waste Reduction/Waste Minimization Program
P.O. Box 10385
Jackson, MS 39289-0385
(601) 961-5171

Missouri

Waste Management Program (WMP)
Becky Shannon, Pollution Prevention Coordinator
Hazardous Waste Program
Division of Environmental Quality
Missouri Department of Natural Resources
205 Jefferson Street
P.O. Box 176
Jefferson City, MO 65102
(314) 751-3176

Montana

Bill Potts
Solid and Hazardous Waste Bureau
Montana Department of Health and Environmental
Sciences
Cogswell Building
Helena, MT 59620
(406) 444-2821

Nebraska

Hazardous Waste Section
Teri Swarts, Waste Minimization Coordinator
Hazardous Waste Section
Nebraska Department of Environmental Control
301 Centennial Mall South
P.O. Box 98922
Lincoln, NE 68509
(402) 471-4217

Nevada

Business Environmental Program
Kevin Dick, Manager
Business Environmental Program
Nevada Small Business Development Center
University of Nevada - Reno
Reno, NV 89557-0100
(702) 784-1717

New Hampshire

New Hampshire Pollution Prevention Program
Vincent R. Perelli
Waste Management Division
New Hampshire Department of Environmental
Service
6 Hazen Street
Concord, NH 03301-6509
(603) 271-2902

New Jersey

New Jersey Office of Pollution Prevention
Jeanne Herb, Director
Office of Pollution Prevention
New Jersey Department of Environmental Protection
CN-402
401 East State Street
Trenton, NJ 08625
(609) 777-0518

New Mexico

Municipal Water Pollution Prevention Program
Alex Puglisi, Program Manager
Municipal Water Pollution Prevention Program
Facility Operations Section, Surface Water Quality
Bureau
New Mexico Environment Department
1190 St. Francis Drive
P.O. Box 26110
Santa Fe, NM 87502
(505) 827-2804

New York

Bureau of Pollution Prevention
John Lanotti, Director
Bureau of Pollution Prevention
Division of Hazardous Substances Regulation and the
Division of Solid Waste
New York State Department of Environmental
Conservation
50 Wolf Road
Albany, NY 12233-7253
(518) 457-7267

New York

Erie County Office of Pollution Prevention (ECOPP)
Thomas Hersey, Pollution Prevention Coordinator
Erie County Office of Pollution Prevention
Erie County Office Building
95 Franklin Street
Buffalo, NY 14202
(716) 858-6231

North Carolina

Pollution Prevention Program
Gary Hunt, Director
Pollution Prevention Program
Office of Waste Reduction
North Carolina Department of Environment, Health,
    and Natural Resources
P.O. Box 27687
Raleigh, NC 27611-7687
(919) 5710-4100

EPA Research Center For Waste Minimization
Dr. Michael Overcash
Dr. Cliff Kaufman
Center for Waste Minimization and Management
North Carolina State University
Box 7905
Raleigh, NC 27695-7905
(919) 515-2325

North Dakota

(No formal State program to date)
Neil Knatterud
Teri Lunde
Division of Waste Management
North Dakota Department of Health and
    Consolidated Laboratories
P.O. Box 5520
1200 Missouri Ave., Room 302
Bismarck, ND 58502-5520
(703) 221-5166

Ohio

Ohio Environmental Protection Agency
Andrea Futrell
Roger Hannahs
Michael W. Kelley
Anthony Sasson
Pollution Prevention Section
Division of Hazardous Waste Management
Ohio Environmental Protection Agency
P.O. Box 1049
1800 Watermark Drive
Columbus, OH 43266-0149
(614) 644-3020

Ohio

Ohio Technology Transfer Organization (OTTO)
Jeff Shick, State Coordinator
Ohio Technology Transfer Organization
Ohio Department of Development
77 South High Street, 26th Floor
Columbus, Ohio 43255-0330
(614) 644-4286

Dan Berglund
Ohio's Thomas Edison Program
77 South High Street, 26th Floor
Columbus, OH 43215
(614) 466-3887

Oklahoma

Environmental Quality Council
Ellen Bussert
Mary Jane Calvey
Environment Health Administration - 0200
1000 North East 10th St.
Oklahoma State Department of Health
Oklahoma City, OK 73117-1299

Pollution Prevention Technical Assistance Program
Chris Varga
Hazardous Waste Management Service, 0205
Oklahoma State Department of Health
1000 Northeast 10th Street
Oklahoma City, OK 73117-1299
(405) 271-7047

Oregon

Waste Reduction Assistance Program (WRAP)
Roy W. Brower, Manager
David Rozell, Pollution Prevention Specialist
Phil Berry, Pollution Prevention Specialist
Hazardous Waste Reduction and Technical Assistance
    Program
Hazardous and Solid Waste Division
Oregon Department of Environmental Quality
811 SW Sixth Avenue
Portland, OR 97204
(503) 229-6585

Pennsylvania

Department of Environmental Resources
Keith Kerns, Chief
Greg Harder
Division of Waste Minimization and Planning
Pennsylvania Department of Environmental
    Resources
P.O. Box 2063
Harrisburg, PA 17120
(717) 772-2724

Pennsylvania Technical Assistance Program
    (PENNTAP)
Jack Gido, Director
PennTap
248 Calder Way, Suite 306
University Park, PA 16801
(814) 865-1914

Rhode Island

Hazardous Waste Reduction Program
Richard Enander, Principal Environmental Scientist
Eugene Pepper, Senior Environmental Planner
Hazardous Waste Reduction Section
Office of Environmental Coordination
Rhode Island Department of Environmental
    Management
83 Park Street
Providence, RI 02903
(401) 277-3434

South Carolina

Center for Waste Minimization
Jeffrey DeBossonet, Manager
Center for Waste Minimization
South Carolina Department of Health and
    Environmental Control
2600 Bull Street
Columbia, SC 29201
(802) 734-4715

South Dakota

Waste Management Program
Vonnie Kallmeyn
Office of Waste Management
Division of Environmental Regulations
South Dakota Department of Environmental and
    Natural Resources
319 S. Coteau
c/o 500 E. Capitol Avenue
Pierre, SD 57501
(605) 773-3153

South Dakota

Waste Management Program
Steve Pirner, Division Director
Division of Environmental Regulations
South Dakota Department of Environment and
    Natural Resources
Joe Foss Building
523 E. Capitol Avenue
Pierre, SD 57501-3181
(605) 773-3153

Tennessee

Department of Health and Environment
James Ault
Bureau of Environment
Tennessee Department of Health and Environment
150 9th Avenue, North
Nashville, TN 37219-3657
(615) 742-6547

Texas

Texas Water Commission
Priscilla Seymour, Ph.D.
Richard Craig
Robert C. Steckly
Office of Pollution Prevention and Conservation
Waste Minimization Unit
Taxas Water Commission P.O. Box 13087, Capitol
    Station
Austin, TX 78711-3087
(512) 463-7761

Utah

Department of Environmental Quality
Rusty Lundberg, Chief
Sonja Wallace, Pollution Prevention Coordinator
Planning and Program Section
Division of Solid and Hazardous Waste Management
Department of Environmental Quality
288 North 1460 West Street
Salt Lake City, UT 84114-4880
(801) 538-6170

Vermont

Waste Minimization Program
Paul Maskowitz, Chief
Gary Gulk
Recycling and Resource Conservation Section
Vermont Department of Environmental Conservation
103 South Main Street
Waterbury, VT 05676
(802) 244-7831

Virginia

Waste Minimization Program
Sharon Kenneally-Baxter, Director
Waste Minimization Program
Virginia Department of Waste Management
Monroe Building, 11th Floor
101 N. 14th Street
Richmond, VA 23219
(804) 371-8716

Washington

Waste Reduction, Recycling and Litter Control
    Program
Stan Springer
Joy St. Morgan
Waste Reduction, Recycling and Litter Control
    Program
Washington Department of Ecology
Mail Stop PV-11
Olympia, WA 98504-8711
(206) 428-7541

West Virginia

Pollution Prevention and Open Dump Program
    (PPOD)
Michael Dorsey, Assistant Chief
Compliance Monitoring and Enforcement Section
Waste Management Section
West Virginia Division of Natural Resources
1356 Hansford Street
Charleston, WV 25301
(304) 348-5989

Wisconsin

Department of Natural Resources
Lynn Persson, Hazardous Waste Reduction and
    Recycling Coordinator
Kate Cooper, Assistant Recycling Coordinator
Bureau of Solid and Hazardous Waste Management
Wisconsin Department of Natural Resources
P.O. Box 7921 (SW/3)
Madison, WI 53707-7921
(608) 267-3763

Wyoming

Department of Environmental Quality
David Finley, Manager
Solid Waste Management Program
Wyoming Department of Environmental Quality
122 West 25th Street
Herschler Building
Cheyenne, WY 82002
(307) 777-7752

Source: These names and addresses are provided by the Environmental Protection Agency in *Pollution Prevention Resources and Training Opportunities in 1992* (EPA/560/8-92-002) January 1992.

C Overview of Known Health Effects of Toxics Release Inventory Chemicals

Classifications of Health and Environmental Effects of TRI Chemicals

Tables C-1 and C-2, on the following pages, include health and environmental assessments for the chemicals on the Toxic Release Inventory (TRI) list of hazardous substances (as of early 1993). A "box" indicates that "the available data support a concern" that the chemical is or may constitute a threat. In other words, a "box" means that studies to date provide some convincing evidence that a chemical is or may be associated with a health or environmental threat. A blank indicates either no concern, no reported effects, or no testing of the chemical to date. The classifications of potential health and environmental effects are:

Carcinogenicity. The potential to cause cancer.

Heritable genetic and chromosomal mutations. Mutations in human reproductive cells that can be passed from generation to generation.

Developmental toxicity. Damages development in the womb or after birth, causing such problems as structural defects, prenatal death, learning disorders, and growth retardation.

Reproductive toxicity. Damages ability to reproduce, causing such problems as sterility, inability to have sex, and inability to produce milk.

Acute toxicity. Causes death after short-term exposure.

Chronic toxicity. Causes adverse effects (other than cancer) after long-term exposure, such as damage to the kidneys, lungs, liver, or bone.

Neurotoxicity. Damages the central or peripheral nervous system after long-term exposure.

Environmental toxicity. Seriously damages the environment and harms wildlife or plants.

Bioaccumulation. Retention or accumulation of a chemical in an organism.

Persistence in the environment. The tendency of a chemical to remain for a long time in the air, water, or land.

(These definitions were derived from: *Support Documentation for the SARA Title III, Sections 313/312 Toxicity Matrix.* Prepared for US EPA by Clement Associates, Inc., Fairfax, VA., August 1988.)

Hazardous Substance Fact Sheets

The Right-to-Know program in the New Jersey Department of Health (DOH) has prepared more than 1,000 Hazardous Substance Fact Sheets. They contain all of the 300 or so chemicals on the TRI list, and several others. In addition, 230 of the fact sheets have been translated into Spanish.

The Hazardous Substance Fact Sheets contain straightforward, easy to understand information about the chemicals. Thus, DOH's Hazardous Substance Fact Sheets (if they are available for the substance you are interested in) are probably more useful to you than Material Safety Data Sheets (MSDS) (discussed in "Surveying Your Local Plant"). While the MSDSs are important, they often contain technical jargon that is slightly more difficult for a layperson to understand.

Each fact sheet contains the following information:

Chemical common name

Chemical Abstract Services (CAS) number

Department of Transportation (DOT) number

Hazard summary

Chemical identification

How to determine exposure

Workplace exposure limits

Ways of reducing exposure

Health hazard information

Medical testing

Workplace controls and practices

Personal protective equipment

Commonly asked questions and answers

Emergency Information:

Fire hazards

Spills and emergencies

Handling and storage

First aid

Physical Data

Other commonly used names

The fact sheets are available from:

The New Jersey Department of Health
Right-to-Know Program, CN 368
Trenton, New Jersey 08625-0368
(609) 984-2202

DOH can send you a complete list of available hazardous substance fact sheets (in English and Spanish), including how to obtain the fact sheets through various databases, and other general information about the program.

Your state TRI contacts (listed in Appendix A) may also have the New Jersey Hazardous Substance Fact Sheets.

Appendix C — Alphabetical Chemical Lists

Table C-1

Chemical Name	CAS Number	Carcinogenicity	Heritable genetic and chromosomal mutations	Developmental toxicity	Reproductive toxicity	Acute toxicity	Chronic (system) toxicity	Neurotoxicity	Environmental toxicity	Bioaccumulation	Persistence in the environment
Acetaldehyde	75-07-0	■				■			■		
Acetamide	60-35-5	■									■
Acetone	67-64-1						■		■		■
Acetonitrile	75-05-8			■	■	■	■				■
2-Acetylaminofluorene	53-96-3	■	■								
Acrolein	107-02-8					■	■		■	■	
Acrylamide	79-06-1	■	■		■	■	■	■	■		
Acrylic acid	79-10-7					■	■		■		
Acrylonitrile	107-13-1	■		■	■	■	■		■		
Aldrin	309-00-2	■		■	■	■	■	■	■	■	
Allyl alcohol	107-18-6					■	■				
Allyl chloride	107-05-1	■		■		■	■	■	■		
alpha-Naphthylamine	134-32-7	■									
Aluminum (fume or dust)	7429-90-5	■									
1-Amino-2-methylanthraquinone	82-28-0	■									
4-Aminoazobenzene	60-09-3	■									
4-Aminobiphenyl	92-67-1	■				■					
Ammonia	7664-41-7					■	■		■		
Ammonium nitrate (solution)	6484-52-2						■				
Ammonium sulfate (solution)	7783-20-2								■		
Aniline	62-53-3	■				■	■		■		■
o-Anisidine hydrochloride	134-29-2	■					■	■			
o-Anisidine	90-04-0	■					■	■			
p-Anisidine	104-94-9										
Anthracene	120-12-7						■		■	■	
Antimony	7440-36-0				■	■	■			■	
Arsenic	7440-38-2	■		■		■	■	■	■	■	
Asbestos (friable)	1332-21-4	■					■				

Chemical Name	CAS Number	Carcinogenicity	Heritable genetic and chromosomal mutations	Developmental toxicity	Reproductive toxicity	Acute toxicity	Chronic (system) toxicity	Neurotoxicity	Environmental toxicity	Bioaccumulation	Persistence in the environment
Auramine	492-80-8	■									
Barium	7440-39-3			■			■				
Benzal chloride	98-87-3	■				■					
Benzamide	55-21-0										
Benzene	71-43-2	■		■	■		■		■		
Benzidine	92-87-5	■				■	■		■		
p-Benzoquinone	106-51-4					■			■		
Benzotrichloride	98-07-7	■		■			■				
Benzoyl chloride	98-88-4			■		■	■		■		
Benzoyl peroxide	94-36-0										
Benzyl chloride	100-44-7	■		■		■	■	■	■		
Beryllium	7440-41-7	■				■	■				
Biphenyl	92-52-4			■			■		■		
Bis(2-ethylhexyl) adipate	103-23-1		■								
Bromochlorodifluoromethane (Halon 1221)	353-59-3					■					
Bromoform	75-25-2						■	■	■		
Bromotrifluoromethane (Halon 1301)	75-63-8							■			
1,3-Butadiene	106-99-0	■		■	■		■	■	■		
1-Butanol	71-36-3						■				
Butyl acrylate	141-32-2								■		
sec-Butyl alcohol	78-92-2										
tert-Butyl alcohol	75-65-0										
Butyl benzyl phthalate	85-68-7						■		■		
1,2-Butylene oxide	106-88-7										
Butyraldehyde	123-72-8								■		
Cadmium	7440-43-9	■		■	■	■	■		■	■	
Calcium cyanamide	156-62-7					■	■				
Captan	133-06-2	■	■	■	■	■	■		■		■
Carbaryl	63-25-2			■	■	■	■	■	■	■	■
Carbon disulfide	75-15-0			■	■		■	■	■		■

Chemical Name	CAS Number	Carcinogenicity	Heritable genetic and chromosomal mutations	Developmental toxicity	Reproductive toxicity	Acute toxicity	Chronic (system) toxicity	Neurotoxicity	Environmental toxicity	Bioaccumulation	Persistence in the environment
Carbon tetrachloride	56-23-5	■		■		■	■	■	■		■
Carbonyl sulfide	463-58-1						■				
Catechol	120-80-9					■			■		
Chloramben	133-90-4	■									
Chlordane	57-74-9	■		■	■	■	■	■		■	■
Chlorine	7782-50-5					■	■		■		
Chlorine dioxide	10049-04-4			■	■						
Chloroacetic acid	79-11-8					■					
2-Chloroacetophenone	532-27-4					■					
Chlorobenzene	108-90-7				■		■		■		
Chlorobenzilate	510-15-6	■			■		■		■		■
Chloroform	67-66-3	■		■	■	■	■		■		■
Chloromethyl methyl ether	107-30-2	■				■					
Chloroprene	126-99-8		■	■	■	■	■	■			
Chlorothalonil	1897-45-6	■				■	■	■	■		■
Chromium	7440-47-3	■				■	■		■		
Cobalt	7440-48-4						■				
Copper	7440-50-8			■	■				■	■	
Creosote	8001-58-9										
p-Cresidine	120-71-8	■									
m-Cresol	108-39-4					■			■		
o-Cresol	95-48-7					■	■	■	■		
p-Cresol	106-44-5					■			■		
Cresol (mixed isomers)	1319-77-3					■	■		■		
Cumene hydroperoxide	80-15-9					■			■		
Cupferron	135-20-6	■				■					
Cyanide compounds	57-12-5			■		■	■	■	■		
Cyclohexane	110-82-7								■		
C.I. Acid Green 3	4680-78-8	■									
C.I. Basic Green 4	569-64-2					■			■		

Chemical Name	CAS Number	Carcinogenicity	Heritable genetic and chromosomal mutations	Developmental toxicity	Reproductive toxicity	Acute toxicity	Chronic (system) toxicity	Neurotoxicity	Environmental toxicity	Bioaccumulation	Persistence in the environment
C.I. Basic Red 1	989-38-8	■									
C.I. Direct Black 38	1937-37-7	■					■				
C.I. Direct Blue 6	2602-46-2	■	■				■				
C.I. Direct Brown 95	16071-86-6	■					■				
C.I. Disperse Yellow 3	2832-40-8	■									
C.I. Food Red 15	81-88-9	■					■				
C.I. Food Red 5	3761-53-3	■									
C.I. Solvent Orange 7	3118-97-6										
C.I. Solvent Yellow 14	842-07-9	■									
C.I. Solvent Yellow 3	97-56-3	■								■	
C.I. Vat Yellow 4	128-66-5	■									
2,4-D	94-75-7			■	■	■	■		■		
Decabromodiphenyl oxide	1163-19-5	■		■			■			■	■
Diallate	2303-16-4	■			■	■			■		■
2,4-Diaminoanisole	615-05-4	■									
2,4-Diaminoanisole sulfate	39156-41-7	■									
4,4'-Diaminodiphenyl ether	101-80-4	■					■				
2,4-Diaminotoluene	95-80-7	■	■								
Diazomethane	334-88-3	■									
Dibenzofuran	132-64-9										
1,2-Dibromo-3-chloropropane	96-12-8	■		■	■	■	■		■		■
Dibromotetrafluoromethand (Halon 2402)	124-73-2					■					
Dibutyl phthalate	84-74-2			■	■	■	■		■	■	
Dichlorobenzene	25321-22-6	■					■		■	■	
m-Dichlorobenzene	541-73-1								■		
o-Dichlorobenzene	95-50-1					■	■		■	■	
p-Dichlorobenzene	106-46-7	■					■		■	■	■
3,3'-Dichlorobenzidine	91-94-1	■									
Dichlorobromomethane	75-27-4										■

Chemical Name	CAS Number	Carcinogenicity	Heritable genetic and chromosomal mutations	Developmental toxicity	Reproductive toxicity	Acute toxicity	Chronic (system) toxicity	Neurotoxicity	Environmental toxicity	Bioaccumulation	Persistence in the environment
Dichlorodifluoromethane (CFC-12)	75-71-8					■	■	■			
Dichloroethyl ether	111-44-4	■				■	■				■
1,1-Dichloroethylene	75-35-4	■		■	■	■	■		■		
Dichloroisopropyl ether	108-60-1	■				■	■				
Dichloromethyl ether	542-88-1	■				■					
2,4-Dichlorophenol	120-83-2								■		
1,3-Dichloropropylene	542-75-6	■				■	■		■		■
2,3-Dichloropropene	78-88-6						■				
Dichlorotetrafluoroethane (CFC-114)	76-14-2					■		■			
Dichlorvos	62-73-7			■	■	■			■		■
Dicofol	115-32-2	■				■			■		
1:2,3:4-Diepoxybutane	1464-53-5	■	■			■					
Diethanolamine	111-42-2	■							■		
Diethyl sulfate	64-67-5	■	■								
Diethylhexyl phthalate	117-81-7	■	■	■	■	■	■		■	■	
Diethylphthalate	84-66-2								■		
3,3'-Dimethoxybenzidine	119-90-4	■					■				
Dimethyl sulfate	77-78-1	■	■			■	■	■	■		■
4-Dimethylaminoazobenzene	60-11-7	■				■				■	
N,N-Dimethylaniline	121-69-7						■	■			
3,3'-Dimethylbenzidine	119-93-7	■									
Dimethylcarbomyl chloride	79-44-7	■				■					
1,1-Dimethylhydrazine	57-14-7	■				■	■		■		■
2,4-Dimethylphenol	105-67-9								■		
Dimethylphthalate	131-11-3					■			■		
M-Dinitrobenzene	99-65-0				■	■	■				
O-Dinitrobenzene	528-29-0				■		■				
P-Dinitrobenzene	100-25-4				■		■				
2,4-Dinitrophenol	51-28-5			■	■	■	■		■		
Dinitrotoluene (mixed isomers)	25321-14-6										

Chemical Name	CAS Number	Carcinogenicity	Heritable genetic and chromosomal mutations	Developmental toxicity	Reproductive toxicity	Acute toxicity	Chronic (system) toxicity	Neurotoxicity	Environmental toxicity	Bioaccumulation	Persistence in the environment
2,4-Dinitrotoluene	121-14-2	■			■		■	■	■		
2,6-Dinitrotoluene	606-20-2	■				■	■	■	■		
4,6-Dinitro-o-cresol	534-52-1					■	■		■		■
Dioxane	123-91-1	■									
1,2-Diphenylhydrazine	122-66-7	■					■		■		
Di-n-octylphthalate	117-84-0					■	■			■	
Di-n-propylnitrosamine	621-64-7	■									■
Epichlorohydrin	106-89-8	■	■		■	■	■	■	■		■
Ethyl acrylate	140-88-5	■		■			■		■		■
Ethyl carbamate (urethane)	51-79-6	■	■								
Ethyl chloride	75-00-3										
Ethyl chloroformate	541-41-3					■					
Ethylbenzene	100-41-4			■	■		■		■		■
Ethylene	74-85-1						■				
Ethylene dibromide	106-93-4	■	■	■	■	■	■		■		■
Ethylene dichloride	107-06-2	■	■	■	■		■				■
Ethylene glycol	107-21-1						■				
Ethylene glycol monoethyl ether	110-80-5			■	■		■				
Ethylene oxide	75-21-8	■	■	■	■	■	■	■	■		■
Ethyleneimine	151-56-4	■	■		■	■	■		■		■
Ethylenethiourea	96-45-7	■	■	■	■		■				■
Fluometuron	2164-17-2	■					■		■		
Formaldehyde	50-00-0	■	■		■	■	■		■		
Freon 113	76-13-1					■	■				
Heptachlor	76-44-8	■			■	■	■	■	■	■	■
Hexachlorobenzene	118-74-1	■		■	■	■	■	■	■	■	■
Hexachlorobutadiene	87-68-3	■			■	■	■	■	■	■	■
Hexachlorocyclopentadiene	77-47-4			■	■	■	■	■	■	■	■
Hexachloroethane	67-72-1	■		■	■	■	■		■		
Hexachloronaphthalene	1335-87-1										

Chemical Name	CAS Number	Carcinogenicity	Heritable genetic and chromosomal mutations	Developmental toxicity	Reproductive toxicity	Acute toxicity	Chronic (system) toxicity	Neurotoxicity	Environmental toxicity	Bioaccumulation	Persistence in the environment
Hexamethylphosphoramide	680-31-9	■	■				■				
Hydrazine	302-01-2	■				■	■		■		
Hydrazine sulfate	10034-93-2	■				■		■			
Hydrochloric acid	7647-01-0					■	■				
Hydrogen cyanide	74-90-8					■	■		■		
Hydrogen fluoride	7664-39-3		■	■	■	■	■				
Hydroquinone	123-31-9					■	■		■		
Isobutyraldehyde	78-84-2						■				
Isopropyl alcohol	67-63-0	■					■	■			
4,4'-Isopropylidenediphenol	80-05-7										
Isosafrole	120-58-1		■								
Lead	7439-92-1	■		■	■		■	■	■	■	■
Lindane	58-89-9	■		■	■	■	■		■	■	■
Maleic anhydride	108-31-6					■	■		■		
Maneb	1247-38-2			■	■	■	■	■	■		■
Manganese	7439-96-5				■	■	■	■			■
MBI	101-68-8										
Mercury	7439-97-6			■	■	■	■	■	■	■	■
Methanol	67-56-1					■		■			
Methoxychlor	72-43-5			■	■	■	■	■	■	■	■
2-Methoxyethanol	109-86-4			■	■		■				
Methyl acrylate	96-33-3					■	■		■		
Methyl bromide	74-83-9	■				■	■	■	■		■
Methyl chloride	74-87-3			■	■		■				■
1-Methyl ethyl benzene (Cumene)	98-82-8						■	■	■		
Methyl ethyl ketone (MEK)	78-93-3			■	■		■				■
Methyl hydrazine	60-34-4					■	■		■		■
Methyl iodide	74-88-4	■				■		■			■
Methyl isobutyl ketone	108-10-1						■	■			
Methyl isocyanate	624-83-9			■	■	■					■

Chemical Name	CAS Number	Carcinogenicity	Heritable genetic and chromosomal mutations	Developmental toxicity	Reproductive toxicity	Acute toxicity	Chronic (system) toxicity	Neurotoxicity	Environmental toxicity	Bioaccumulation	Persistence in the environment
Methyl methacrylate	80-62-6			■	■		■				
Methyl tert-butyl ether	1634-04-4					■					
Methylene bromide	74-95-3					■		■			
Methylene chloride	75-09-2	■					■				■
4,4'-Methylenebis (2-chloroaniline)	101-14-4	■									
4,4'-Methylenebis (N,N-dimethyl)	101-61-1	■					■				
4,4'-Methylenedianiline	101-77-9	■				■	■				
Michler's ketone	90-94-8	■				■					
Molybdenum trioxide	1313-27-5					■	■	■			
Monochloropentafluoroethane (CFC-115)	76-15-3							■			
Mustard gas	505-60-2	■	■			■					
Naphthalene	91-20-3			■			■		■	■	
beta-Naphthylamine	91-59-8	■									
Nickel	7440-02-0	■		■	■	■	■		■		
Nitric acid	7697-37-2					■					
Nitrilotriacetic acid	139-13-9	■					■				
Nitrobenzene	98-95-3				■	■	■		■		
4-Nitrobiphenyl	92-93-3	■									
Nitrofen	1836-75-5	■		■	■		■	■	■		
Nitrogen mustard	51-75-2	■	■	■	■	■					
Nitroglycerin	55-63-0					■			■		
2-Nitrophenol	88-75-5						■	■	■		
p-Nitrophenol	100-02-7					■	■	■	■		
2-Nitropropane	79-46-9	■		■	■	■	■				
N-Nitrosodiethylamine	55-18-5	■	■	■	■	■					
N-Nitrosodimethylamine	62-75-9	■	■			■					
N-Nitrosodiphenylamine	86-30-6	■					■		■		
p-Nitrosodiphenylamine	156-10-5	■									
N-Nitrosodi-n-butylamine	924-16-3	■									

Chemical Name	CAS Number	Carcinogenicity	Heritable genetic and chromosomal mutations	Developmental toxicity	Reproductive toxicity	Acute toxicity	Chronic (system) toxicity	Neurotoxicity	Environmental toxicity	Bioaccumulation	Persistence in the environment
N-Nitrosomethylvinylamine	4549-40-0	■	■			■					
N-Nitrosomorpholine	59-89-2	■	■	■			■				
N-Nitrosonornicotine	16543-55-8	■									
N-Nitrosopiperidine	100-75-4	■	■			■	■				
N-Nitroso-N-ethyl urea	759-73-9	■	■	■	■						
N-Nitroso-N-methylurea	684-93-5	■	■	■		■					
5-Nitro-o-anisidine	99-59-2	■				■					
Octachloronaphthalene	2234-13-1				■				■		
Osmium tetroxide	20816-12-0					■					
Parathion	56-38-2			■	■	■	■		■		
PCBs	1336-36-3	■		■	■	■	■				
Pentachloronitrobenzene (PCNB)	82-68-8	■		■	■	■			■	■	
Pentachlorophenol	87-86-5			■	■	■	■		■	■	■
Peracetic acid	79-21-0					■	■				
Phenol	108-95-2				■	■	■		■	■	■
p-Phenylenediamine	106-50-3		■			■	■		■		
2-Phenylphenol	90-43-7	■		■	■		■				
Phosgene	75-44-5					■					■
Phosphoric acid	7664-38-2										
Phosphorous (yellow or white)	7723-14-0				■	■	■		■		
Phthalic anhydride	85-44-9						■	■	■		
Picric acid	88-89-1								■		
1,3-Propane sultone	1120-71-4	■									
beta-Propiolactone	57-57-8	■	■			■					
Propionaldehyde	123-38-6						■		■		
Propxur	114-26-1					■	■				
Propylene	115-07-1										
Propylene dichloride	78-87-5	■							■		
Propylene oxide	75-56-9	■	■	■	■	■	■	■	■		■
Propyleneimine	75-55-8	■				■					

Table C-1 — Alphabetical Listing

Chemical Name	CAS Number	Carcinogenicity	Heritable genetic and chromosomal mutations	Developmental toxicity	Reproductive toxicity	Acute toxicity	Chronic (system) toxicity	Neurotoxicity	Environmental toxicity	Bioaccumulation	Persistence in the environment
Pyridine	110-86-1					■	■		■		■
Quinoline	91-22-5					■	■		■		■
Saccharin	81-07-2	■	■	■	■	■					
Safrole	94-59-7	■			■		■				
Selenium	7782-49-2			■	■		■		■		
Styrene	100-42-5	■	■	■			■				
Styrene oxide	96-09-3	■		■	■						
Sulfuric acid	7664-93-9					■	■		■		
1,1,2,2-Tetrachloroethane	79-34-5	■		■			■	■	■		■
Tetrachloroethylene	127-18-4	■		■	■		■	■	■		■
Tetrachlorvinphos	961-11-5	■				■	■		■		■
Thallium	7440-28-0										
Thioacetamide	62-55-5	■	■				■				
4,4'-Thiodianiline	139-65-1	■			■						
Thiourea	62-56-6	■		■	■	■					
Thorium dioxide	1314-20-1	■									
Titanium tetrachloride	7550-45-0					■					
Toluene	108-88-3			■	■				■		
Toluenediamine	25376-45-8	■	■				■				
Toluene diisocyanate (mixed isomers)	26471-62-5					■	■		■		
Toluene-2,4-diisocyanate	584-84-9	■				■			■		
Toluene-2,6-diisocyanate	91-08-7					■	■				
o-Toluidine hydrochloride	636-21-5	■									
o-Toluidine	95-53-4	■					■		■		
Toxaphene	8001-35-2	■		■	■	■	■	■	■	■	■
Triaziquone	68-76-8	■	■		■						
Trichlorfon	52-68-6			■	■	■	■		■		■
1,2,4-Trichlorobenzene	120-82-1			■			■		■	■	■
1,1,1-Trichloroethane	71-55-6			■	■	■	■		■		■

Chemical Name	CAS Number	Carcinogenicity	Heritable genetic and chromosomal mutations	Developmental toxicity	Reproductive toxicity	Acute toxicity	Chronic (system) toxicity	Neurotoxicity	Environmental toxicity	Bioaccumulation	Persistence in the environment
1,1,2-Trichloroethane	79-00-5	■					■		■		■
Trichloroethylene	79-01-6	■		■	■		■	■			■
Trichlorofluoromethane (CFC-11)	75-69-4					■	■	■			
2,4,5-Trichlorophenol	95-95-4					■	■		■		
2,4,6-Trichlorophenol	88-06-2	■							■		■
Trifluralin	1582-09-8	■		■	■		■		■	■	
1,2,4-Trimethylbenzene	95-63-6					■			■		
Tris (2,3-dibromopropyl) phosphate	126-72-7	■	■	■	■						
Vanadium (fume or dust)	7440-62-2										
Vinyl acetate	108-05-4						■		■		
Vinyl bromide	593-60-2	■					■				
Vinyl chloride	75-01-4	■	■	■	■	■	■				■
m-Xylene	108-38-3			■		■	■		■		■
o-Xylene	95-47-6			■	■		■		■		■
p-Xylene	106-42-3			■	■	■			■		■
Xylene (mixed isomers)	1330-20-7			■	■		■		■		
2,6-Xylidine	87-62-7	■					■				
Zinc (fume or dust)	7440-66-6			■	■		■		■	■	
Zineb	12122-67-7				■		■		■		■

Chemical Name	CAS Number	Carcinogenicity	Heritable genetic and chromosomal mutations	Developmental toxicity	Reproductive toxicity	Acute toxicity	Chronic (system) toxicity	Neurotoxicity	Environmental toxicity	Bioaccumulation	Persistence in the environment
Formaldehyde	50-00-0	■	■		■	■	■		■		
2,4-Dinitrophenol	51-28-5			■	■	■	■		■		
Nitrogen mustard	51-75-2	■	■	■	■	■					
Ethyl carbamate (urethane)	51-79-6	■	■								
Trichlorfon	52-68-6			■	■	■	■		■		■
2-Acetylaminofluorene	53-96-3	■	■								
N-Nitrosodiethylamine	55-18-5	■	■	■	■	■					
Benzamide	55-21-0										
Nitroglycerin	55-63-0					■			■		
Carbon tetrachloride	56-23-5	■		■		■	■	■	■		■
Parathion	56-38-2			■	■	■	■		■		
Cyanide compounds	57-12-5			■		■	■	■	■		
1,1-Dimethylhydrazine	57-14-7	■				■	■		■		■
beta-Propiolactone	57-57-8	■	■			■					
Chlordane	57-74-9	■		■	■	■	■	■		■	■
Lindane	58-89-9	■		■	■	■	■		■	■	■
N-Nitrosomorpholine	59-89-2	■	■	■			■				
4-Aminoazobenzene	60-09-3	■									
4-Dimethylaminoazobenzene	60-11-7	■				■				■	
Methyl hydrazine	60-34-4					■	■		■		■
Acetamide	60-35-5	■									■
Aniline	62-53-3	■				■	■		■		■
Thioacetamide	62-55-5	■	■				■				
Thiourea	62-56-6	■		■	■	■					
Dichlorvos	62-73-7			■	■	■			■		■
N-Nitrosodimethylamine	62-75-9	■	■			■					
Carbaryl	63-25-2			■	■	■	■	■	■	■	■
Diethyl sulfate	64-67-5	■	■								

Table C-2 — Listing by CAS Number

Chemical Name	CAS Number	Carcinogenicity	Heritable genetic and chromosomal mutations	Developmental toxicity	Reproductive toxicity	Acute toxicity	Chronic (system) toxicity	Neurotoxicity	Environmental toxicity	Bioaccumulation	Persistence in the environment
Methanol	67-56-1					■		■			
Isopropyl alcohol	67-63-0	■					■	■			
Acetone	67-64-1						■		■		■
Chloroform	67-66-3	■		■	■	■	■		■		■
Hexachloroethane	67-72-1	■		■	■	■	■		■		
Triaziquone	68-76-8	■	■		■						
1-Butanol	71-36-3						■				
Benzene	71-43-2	■		■	■		■		■		
1,1,1-Trichloroethane	71-55-6			■	■	■	■		■		■
Methoxychlor	72-43-5			■	■	■	■	■	■	■	■
Methyl bromide	74-83-9	■				■	■	■	■		■
Ethylene	74-85-1						■				
Methyl chloride	74-87-3			■	■		■				■
Methyl iodide	74-88-4	■				■		■			■
Hydrogen cyanide	74-90-8					■	■		■		
Methylene bromide	74-95-3					■		■			
Ethyl chloride	75-00-3										
Vinyl chloride	75-01-4	■	■	■	■	■	■				■
Acetonitrile	75-05-8			■	■	■	■				■
Acetaldehyde	75-07-0	■				■			■		
Methylene chloride	75-09-2	■					■				■
Carbon disulfide	75-15-0			■	■		■	■	■		■
Ethylene oxide	75-21-8	■	■	■	■	■	■	■	■		■
Bromoform	75-25-2						■	■	■		
Dichlorobromomethane	75-27-4										■
1,1-Dichloroethylene	75-35-4	■		■	■		■		■		
Phosgene	75-44-5					■					■
Propyleneimine	75-55-8	■				■					
Propylene oxide	75-56-9	■	■	■	■	■	■	■	■		■
tert-Butyl alcohol	75-65-0										

Chemical Name	CAS Number	Carcinogenicity	Heritable genetic and chromosomal mutations	Developmental toxicity	Reproductive toxicity	Acute toxicity	Chronic (system) toxicity	Neurotoxicity	Environmental toxicity	Bioaccumulation	Persistence in the environment
Bromotrifluoromethane (Halon 1301)	75-63-8							■			
Trichlorofluoromethane (CFC-11)	75-69-4					■	■	■			
Dichlorodifluoromethane (CFC-12)	75-71-8					■	■	■			
Freon 113	76-13-1					■	■				
Dichlorotetrafluoroethane (CFC-114)	76-14-2					■		■			
Monochloropentafluoroethane (CFC-115)	76-15-3							■			
Heptachlor	76-44-8	■			■	■	■	■	■	■	■
Hexachlorocyclopentadiene	77-47-4			■	■	■	■	■	■	■	■
Dimethyl sulfate	77-78-1	■	■			■	■	■	■		■
Isobutyraldehyde	78-84-2						■				
Propylene dichloride	78-87-5	■							■		
2,3-Dichloropropene	78-88-6						■				
sec-Butyl alcohol	78-92-2										
Methyl ethyl ketone (MEK)	78-93-3			■	■		■	■			■
1,1,2-Trichloroethane	79-00-5	■					■		■		■
Trichloroethylene	79-01-6	■		■	■		■	■			■
Acrylamide	79-06-1	■	■		■	■	■	■	■		
Acrylic acid	79-10-7					■	■		■		
Chloroacetic acid	79-11-8					■					
Peracetic acid	79-21-0					■	■				
1,1,2,2-Tetrachloroethane	79-34-5	■		■			■	■	■		■
Dimethylcarbomyl chloride	79-44-7	■				■					
2-Nitropropane	79-46-9	■		■	■	■	■				
4,4'-Isopropylidenediphenol	80-05-7										
Cumene hydroperoxide	80-15-9					■			■		
Methyl methacrylate	80-62-6			■	■		■				
Saccharin	81-07-2	■	■	■	■	■					
C.I. Food Red 15	81-88-9	■					■				
1-Amino-2-methylanthraquinone	82-28-0	■									
Pentachloronitrobenzene (PCNB)	82-68-8	■		■	■	■			■	■	

Table C-2 — Listing by CAS Number

Chemical Name	CAS Number	Carcinogenicity	Heritable genetic and chromosomal mutations	Developmental toxicity	Reproductive toxicity	Acute toxicity	Chronic (system) toxicity	Neurotoxicity	Environmental toxicity	Bioaccumulation	Persistence in the environment
Diethylphthalate	84-66-2							■			
Dibutyl phthalate	84-74-2			■	■	■	■		■	■	
Phthalic anhydride	85-44-9						■	■	■		
Butyl benzyl phthalate	85-68-7						■		■		
N-Nitrosodiphenylamine	86-30-6	■					■		■		
2,6-Xylidine	87-62-7	■					■				
Hexachlorobutadiene	87-68-3	■		■	■	■	■		■	■	■
Pentachlorophenol	87-86-5			■	■	■	■		■	■	■
2,4,6-Trichlorophenol	88-06-2	■							■		■
2-Nitrophenol	88-75-5						■	■	■		
Picric acid	88-89-1							■			
o-Anisidine	90-04-0	■					■	■			
2-Phenylphenol	90-43-7	■		■	■		■				
Michler's ketone	90-94-8	■				■					
Toluene-2,6-diisocyanate	91-08-7					■	■				
Naphthalene	91-20-3			■			■		■	■	
Quinoline	91-22-5					■	■		■		■
beta-Naphthylamine	91-59-8	■									
3,3' Dichlorobenzidine	91-94-1	■									
Biphenyl	92-52-4			■			■		■		
4-Aminobiphenyl	92-67-1	■				■					
Benzidine	92-87-5	■				■	■		■		
4-Nitrobiphenyl	92-93-3	■									
Benzoyl peroxide	94-36-0										
Safrole	94-59-7	■			■		■				
2,4-D	94-75-7			■	■	■			■		
o-Xylene	95-47-6			■	■		■		■		■
o-Cresol	95-48-7					■	■	■	■		
o-Dichlorobenzene	95-50-1						■		■	■	
o-Toluidine	95-53-4	■					■		■		

Table C-2 — Listing by CAS Number

Chemical Name	CAS Number	Carcinogenicity	Heritable genetic and chromosomal mutations	Developmental toxicity	Reproductive toxicity	Acute toxicity	Chronic (system) toxicity	Neurotoxicity	Environmental toxicity	Bioaccumulation	Persistence in the environment
1,2,4-Trimethylbenzene	95-63-6					■			■		
2,4-Diaminotoluene	95-80-7	■	■								
2,4,5-Trichlorophenol	95-95-4					■	■		■		
Styrene oxide	96-09-3	■		■	■						
1,2-Dibromo-3-chloropropane	96-12-8	■		■	■	■	■		■		■
Methyl acrylate	96-33-3					■	■		■		
Ethylenethiourea	96-45-7	■	■	■	■		■				■
C.I. Solvent Yellow 3	97-56-3	■								■	
Benzotrichloride	98-07-7	■		■			■				
1-Methyl ethyl benzene (Cumene)	98-82-8						■	■	■		
Benzal chloride	98-87-3	■				■					
Benzoyl chloride	98-88-4			■		■	■		■		
Nitrobenzene	98-95-3				■	■	■		■		
5-Nitro-o-anisidine	99-59-2	■				■					
m-Dinitrobenzene	99-65-0				■	■	■				
p-Nitrophenol	100-02-7					■	■	■	■		
p-Dinitrobenzene	100-25-4				■		■				
Ethylbenzene	100-41-4			■	■		■		■		■
Styrene	100-42-5	■	■	■			■				
Benzyl chloride	100-44-7	■		■		■	■	■	■		
N-Nitrosopiperidine	100-75-4	■	■			■	■				
4,4'-Methylenebis (2-chloroaniline)	101-14-4	■									
4,4'-Methylenebis (N,N-dimethyl)	101-61-1	■					■				
MBI	101-68-8										
4,4'-Methylenedianiline	101-77-9	■				■	■				
4,4'-Diaminodiphenyl ether	101-80-4	■					■				
Bis (2-ethylhexyl) adipate	103-23-1		■								
p-Anisidine	104-94-9										
2,4-Dimethylphenol	105-67-9								■		
p-Xylene	106-42-3			■	■	■			■		■

Chemical Name	CAS Number	Carcinogenicity	Heritable genetic and chromosomal mutations	Developmental toxicity	Reproductive toxicity	Acute toxicity	Chronic (system) toxicity	Neurotoxicity	Environmental toxicity	Bioaccumulation	Persistence in the environment
p-Cresol	106-44-5					■			■		
p-Dichlorobenzene	106-46-7	■					■		■	■	■
p-Phenylenediamine	106-50-3		■			■	■		■		
p-Benzoquinone	106-51-4					■			■		
1,2-Butylene oxide	106-89-7										
Epichlorohydrin	106-89-8	■	■		■	■	■	■	■		■
Ethylene dibromide	106-93-4	■	■	■	■	■	■		■		■
1,3-Butadiene	106-99-0	■		■	■		■	■	■		
Acrolein	107-02-8					■	■		■		■
Allyl chloride	107-05-1	■		■		■	■	■	■		
Ethylene dichloride	107-06-2	■	■	■	■		■				■
Acrylonitrile	107-13-1	■		■	■	■	■		■		
Allyl alcohol	107-18-6					■	■				
Ethylene glycol	107-21-1						■				
Chloromethyl methyl ether	107-30-2	■				■					
Vinyl acetate	108-05-4						■		■		
Methyl isobutyl ketone	108-10-1						■	■			
Maleic anhydride	108-31-6					■	■		■		
m-Xylene	108-38-3			■		■	■		■		■
m-Cresol	108-39-4					■			■		
Dichloroisopropyl ether	108-60-1	■				■	■				
Toluene	108-88-3			■	■				■		
Chlorobenzene	108-90-7				■		■		■		
Phenol	108-95-2			■		■	■		■	■	■
2-Methoxyethanol	109-86-4			■	■		■				
Ethylene glycol monoethyl ether	110-80-5			■	■		■				
Cyclohexane	110-82-7								■		
Pyridine	110-86-1					■	■		■		■
Diethanolamine	111-42-2	■				■			■		
Dichloroethyl ether	111-44-4	■				■	■				■

Chemical Name	CAS Number	Carcinogenicity	Heritable genetic and chromosomal mutations	Developmental toxicity	Reproductive toxicity	Acute toxicity	Chronic (system) toxicity	Neurotoxicity	Environmental toxicity	Bioaccumulation	Persistence in the environment
Propxur	114-26-1					■	■				
Propylene	115-07-1										
Dicofol	115-32-2	■				■			■		
2-Aminoanthraquinone	117-79-3	■					■				
Diethylhexyl phthalate	117-81-7	■	■	■	■	■	■		■	■	
Di-n-octylphthalate	117-84-0					■	■			■	
Hexachlorobenzene	118-74-1	■		■	■	■	■	■	■	■	■
3,3'-Dimethoxybenzidine	119-90-4	■					■				
3,3'-Dimethylbenzidine	119-93-7	■									
Anthracene	120-12-7						■		■	■	
Isosafrole	120-58-1		■								
p-Cresidine	120-71-8	■									
Catechol	120-80-9					■			■		
1,2,4-Trichlorobenzene	120-82-1			■			■		■	■	■
2,4-Dichlorophenol	120-83-2								■		
2,4-Dinitrotoluene	121-14-2	■			■		■	■	■		
N,N-Dimethylaniline	121-69-7						■	■			
1,2-Diphenylhydrazine	122-66-7	■					■		■		
Hydroquinone	123-31-9					■	■		■		
Propionaldehyde	123-38-6						■		■		
Butyraldehyde	123-72-8								■		
Dioxane	123-91-1	■									
Dibromotetrafluoroethane (Halon 2402)	124-73-2					■					
Tris (2,3-dibromopropyl) phosphate	126-72-7	■	■	■	■						
Chloroprene	126-99-8		■	■	■	■	■	■			
Tetrachloroethylene	127-18-4	■		■	■		■	■	■		■
C.I. Vat Yellow 4	128-66-5	■									
Dimethylphthalate	131-11-3					■			■		
Dibenzofuran	132-64-9										

Chemical Name	CAS Number	Carcinogenicity	Heritable genetic and chromosomal mutations	Developmental toxicity	Reproductive toxicity	Acute toxicity	Chronic (system) toxicity	Neurotoxicity	Environmental toxicity	Bioaccumulation	Persistence in the environment
Captan	133-06-2	■	■	■	■	■	■		■		■
Chloramben	133-90-4	■									
o-Anisidine hydrochloride	134-29-2	■					■	■			
alpha-Naphthylamine	134-32-7	■									
Cupferron	135-20-6	■				■					
Nitrilotriacetic acid	139-13-9	■					■				
4,4'-Thiodianiline	139-65-1	■			■						
Ethyl acrylate	140-88-5	■		■			■		■		■
Butyl acrylate	141-32-2								■		
Ethyleneimine	151-56-4	■	■		■	■	■		■		■
p-Nitrosodiphenylamine	156-10-5	■									
Calcium cyanamide	156-62-7					■	■				
Hydrazine	302-01-2	■				■	■		■		
Aldrin	309-00-2	■		■	■	■	■	■	■	■	
Diazomethane	334-88-3	■									
Bromochlorodifluoromethane (Halon 1221)	353-59-3					■					
Carbonyl sulfide	463-58-1						■				
Auramine	492-80-8	■									
Mustard gas	505-60-2	■	■			■					
Chlorobenzilate	510-15-6	■			■		■		■		■
o-Dinitrobenzene	528-29-0				■		■				
2-Chloroacetophenone	532-27-4					■					
4,6-Dinitro-o-cresol	534-52-1					■	■		■		■
Ethyl chloroformate	541-41-3					■					
m-Dichlorobenzene	541-73-1								■		
1,3-Dichloropropylene	542-75-6	■				■	■		■		■
Dichloromethyl ether	542-88-1	■				■					
C.I. Basic Green 4	569-64-2					■			■		
Toluene-2,4-diisocyanate	584-84-9	■				■			■		
Vinyl bromide	593-60-2	■					■				

Chemical Name	CAS Number	Carcinogenicity	Heritable genetic and chromosomal mutations	Developmental toxicity	Reproductive toxicity	Acute toxicity	Chronic (system) toxicity	Neurotoxicity	Environmental toxicity	Bioaccumulation	Persistence in the environment
2,6-Dinitrotoluene	606-20-2	■			■	■	■	■			
2,4-Diaminoanisole	615-05-4	■									
Di-n-propylnitrosamine	621-64-7	■									■
Methyl isocyanate	624-83-9			■	■	■					■
o-Toluidine hydrochloride	636-21-5	■									
Hexamethylphosphoramide	680-31-9	■	■				■				
N-Nitroso-N-methylurea	684-93-5	■	■	■		■					
N-Nitroso-N-ethyl urea	759-73-9	■	■	■	■						
C.I. Solvent Yellow 14	842-07-9	■									
N-Nitrosodi-n-butylamine	924-16-3	■									
Tetrachlorvinphos	961-11-5	■				■	■		■		■
C.I. Basic Red 1	989-38-8	■									
1,3-Propane sultone	1120-71-4	■									
Decabromodiphenyl oxide	1163-19-5	■		■			■			■	■
Maneb	1247-38-2			■	■	■	■	■	■		■
Molybdenum trioxide	1313-27-5					■	■	■			
Thorium dioxide	1314-20-1	■									
Cresol (mixed isomers)	1319-77-3					■	■		■		
Xylene (mixed isomers)	1330-20-7			■	■		■		■		
Asbestos (friable)	1332-21-4	■					■				
Hexachloronaphthalene	1335-87-1										
PCBs	1336-36-3	■		■	■	■	■				
1:2,3:4-Diepoxybutane	1464-53-5	■	■			■					
Trifluralin	1582-09-8	■		■	■		■		■	■	
Methyl tert-butyl ether	1634-04-4					■					
Nitrofen	1836-75-5	■		■	■		■	■	■		
Chlorothalonil	1897-45-6	■				■	■	■	■		■
C.I. Direct Black 38	1937-37-7	■					■				
Fluometuron	2164-17-2	■					■		■		
Octachloronaphthalene	2234-13-1				■				■		

Table C-2 — Listing by CAS Number

Chemical Name	CAS Number	Carcinogenicity	Heritable genetic and chromosomal mutations	Developmental toxicity	Reproductive toxicity	Acute toxicity	Chronic (system) toxicity	Neurotoxicity	Environmental toxicity	Bioaccumulation	Persistence in the environment
Diallate	2303-16-4	■			■	■		■			■
C.I. Direct Blue 6	2602-46-2	■		■			■				
C.I. Disperse Yellow 3	2832-40-8	■									
C.I. Solvent Orange 7	3118-97-6										
C.I. Food Red 5	3761-53-3	■									
N-Nitrosomethylvinylamine	4549-40-0	■	■			■					
C.I. Acid Green 3	4680-78-8	■									
Ammonium nitrate (solution)	6484-52-2						■				
Aluminum (fume or dust)	7429-90-5	■									
Lead	7439-92-1	■		■	■		■	■	■	■	■
Manganese	7439-96-5				■	■	■	■			■
Mercury	7439-97-6			■	■	■	■	■	■	■	■
Nickel	7440-02-0	■		■	■	■	■		■		
Thallium	7440-28-0										
Antimony	7440-36-0				■	■	■			■	
Arsenic	7440-38-2	■		■		■	■	■	■	■	
Barium	7440-39-3			■			■				
Beryllium	7440-41-7	■				■	■				
Cadmium	7440-43-9	■		■	■	■	■		■	■	
Chromium	7440-47-3	■				■	■		■		
Cobalt	7440-48-4						■				
Copper	7440-50-8			■	■				■	■	
Vanadium (fume or dust)	7440-62-2										
Zinc (fume or dust)	7440-66-6			■	■		■		■	■	
Titanium tetrachloride	7550-45-0					■					
Hydrochloric acid	7647-01-0					■	■				
Phosphoric acid	7664-38-2										
Hydrogen fluoride	7664-39-3		■	■	■	■	■				
Ammonia	7664-41-7					■	■		■		
Sulfuric acid	7664-93-9					■	■		■		

Chemical Name	CAS Number	Carcinogenicity	Heritable genetic and chromosomal mutations	Developmental toxicity	Reproductive toxicity	Acute toxicity	Chronic (system) toxicity	Neurotoxicity	Environmental toxicity	Bioaccumulation	Persistence in the environment
Nitric acid	7697-37-2					■					
Phosphorous (yellow or white)	7723-14-0				■	■	■		■		
Selenium	7782-49-2			■	■		■		■		
Chlorine	7792-50-5					■	■		■		
Ammonium sulfate (solution)	7783-20-2								■		
Toxaphene	8001-35-2	■		■	■	■	■	■	■	■	■
Creosote	8001-58-9										
Hydrazine sulfate	10034-93-2	■				■			■		
Chlorine dioxide	10049-04-4			■	■						
Zineb	12122-67-7				■		■		■		■
C.I. Direct Brown 95	16071-86-6	■					■				
N-Nitrosonornicotine	16543-55-8	■									
Osmium tetroxide	20816-12-0					■					
Dinitrotoluene (mixed isomers)	25321-14-6										
Dichlorobenzene	25321-22-6	■					■		■	■	
Toluenediamine	25376-45-8	■	■				■				
Toluene diisocyanate (mixed isomers)	26471-62-5					■	■		■		
2,4-Diaminoanisole sulfate	39156-41-7	■									

D US Environmental Protection Agency Reporting Forms

Appendix D contains the following US Environmental Protection Agency forms:

Form R: Toxic Chemical Release Inventory Reporting Form (9 pages)

Emergency and Hazardous Chemical Inventory: Tiers 1 and 2 (3 pages)

Form IC: Identification and Certification (2 pages)

Form GM: Waste Generation and Management (1 page)

	TRI FACILITY ID NUMBER

⊕EPA
United States
Environmental Protection
Agency

FORM R TOXIC CHEMICAL RELEASE
INVENTORY REPORTING FORM

Section 313 of the Emergency Planning and Community Right-to-Know Act of 1986,
also known as Title III of the Superfund Amendments and Reauthorization Act

Toxic Chemical, Category, or Generic Name

WHERE TO SEND COMPLETED FORMS:

1. EPCRA Reporting Center
   P.O. Box 3348
   Merrifield, VA 22116-3348
   ATTN: TOXIC CHEMICAL RELEASE INVENTORY

2. APPROPRIATE STATE OFFICE
   (See instructions in Appendix F)

Enter "X" here if
this is a revision

IMPORTANT: See instructions to determine when "Not
Applicable (NA)" boxes should be checked.

For EPA use only

PART I. FACILITY IDENTIFICATION INFORMATION

SECTION 1.

REPORTING YEAR

19 ____

SECTION 2.   TRADE SECRET INFORMATION

Are you claiming the toxic chemical identified on page 3 trade secret?

2.1 ☐ Yes (Answer question 2.2; Attach substantiation forms) ☐ No (Do not answer 2.2; Go to Section 3)

2.2 If yes in 2.1, is this copy: ☐ Sanitized ☐ Unsanitized

SECTION 3.   CERTIFICATION (Important: Read and sign after completing all form sections.)

I hereby certify that I have reviewed the attached documents and that, to the best of my knowledge and belief, the submitted information is true and complete and that the amounts and values in this report are accurate based on reasonable estimates using data available to the preparers of this report.

Name and official title of owner/operator or senior management official

Signature

Date Signed

SECTION 4. FACILITY IDENTIFICATION

4.1

Facility or Establishment Name

TRI Facility ID Number

Street Address

City

County

State

Zip Code

Mailing Address (if different from street address)

City

State

Zip Code

PUT LABEL HERE

EPA Form 9350-1 (Rev. 12/4/92) - Previous editions are obsolete.

&EPA
United States
Environmental Protection
Agency

EPA FORM R
PART I. FACILITY IDENTIFICATION INFORMATION (CONTINUED)

TRI FACILITY ID NUMBER

Toxic Chemical, Category, or Generic Name

SECTION 4.   FACILITY IDENTIFICATION (Continued)

4.2	This report contains information for: (Important: check only one)	a. ☐ An entire facility b. ☐ Part of a facility
4.3	Technical Contact	Name _______ Telephone Number (include area code) _______
4.4	Public Contact	Name _______ Telephone Number (include area code) _______

4.5 SIC Code (4-digit)

a.	b.	c.	d.	e.	f.

4.6 Latitude and Longitude

	Latitude			Longitude		
	Degrees	Minutes	Seconds	Degrees	Minutes	Seconds

4.7	Dun & Bradstreet Number(s) (9 digits)	a. _______ b. _______
4.8	EPA Identification Number(s) (RCRA I.D. No.) (12 characters)	a. _______ b. _______
4.9	Facility NPDES Permit Number(s) (9 characters)	a. _______ b. _______
4.10	Underground Injection Well Code (UIC) I.D. Number(s) (12 digits)	a. _______ b. _______

SECTION 5. PARENT COMPANY INFORMATION

5.1	Name of Parent Company	☐ NA
5.2	Parent Company's Dun & Bradstreet Number	☐ NA (9 digits)

EPA Form 9350-1 (Rev. 12/4/92) - Previous editions are obsolete.

⊕EPA
United States
Environmental Protection
Agency

EPA FORM R

PART II. CHEMICAL-SPECIFIC INFORMATION

TRI FACILITY ID NUMBER

Toxic Chemical, Category, or Generic Name

SECTION 1. TOXIC CHEMICAL IDENTITY   (Important: DO NOT complete this section if you complete Section 2 below.)

1.1	CAS Number (Important: Enter only one number exactly as it appears on the Section 313 list. Enter category code if reporting a chemical category.)
1.2	Toxic Chemical or Chemical Category Name (Important: Enter only one name exactly as it appears on the Section 313 list.)
1.3	Generic Chemical Name (Important: Complete **only** if Part I, Section 2.1 is checked "yes." Generic Name must be structurally descriptive.)

SECTION 2. MIXTURE COMPONENT IDENTITY   (Important: DO NOT complete this section if you complete Section 1 above.)

2.1	Generic Chemical Name Provided by Supplier (Important: Maximum of 70 characters, including numbers, letters, spaces, and punctuation.)

SECTION 3. ACTIVITIES AND USES OF THE TOXIC CHEMICAL AT THE FACILITY
(Important: Check all that apply.)

3.1 Manufacture the toxic chemical:

a. ☐ Produce
b. ☐ Import

If produce or import:
c. ☐ For on-site use/processing
d. ☐ For sale/distribution
e. ☐ As a byproduct
f. ☐ As an impurity

3.2 Process the toxic chemical:

a. ☐ As a reactant
b. ☐ As a formulation component
c. ☐ As an article component
d. ☐ Repackaging

3.3 Otherwise use the toxic chemical:

a. ☐ As a chemical processing aid
b. ☐ As a manufacturing aid
c. ☐ Ancillary or other use

SECTION 4. MAXIMUM AMOUNT OF THE TOXIC CHEMICAL ON-SITE AT ANY TIME DURING THE CALENDAR YEAR

4.1	☐ (Enter two-digit code from instruction package.)

EPA Form 9350-1(Rev. 12/4/92) - Previous editions are obsolete.

⊕EPA
United States
Environmental Protection
Agency

EPA FORM R
PART II. CHEMICAL-SPECIFIC INFORMATION (CONTINUED)

TRI FACILITY ID NUMBER

Toxic Chemical, Category, or Generic Name

SECTION 5. RELEASES OF THE TOXIC CHEMICAL TO THE ENVIRONMENT ON-SITE

			A. Total Release (pounds/year) (enter range code from instructions or estimate)	B. Basis of Estimate (enter code)	C. % From Stormwater
5.1	Fugitive or non-point air emissions	□ NA			
5.2	Stack or point air emissions	□ NA			
5.3	Discharges to receiving streams or water bodies (enter one name per box)				
5.3.1	Stream or Water Body Name				
5.3.2	Stream or Water Body Name				
5.3.3	Stream or Water Body Name				
5.4	Underground injections on-site	□ NA			
5.5	Releases to land on-site				
5.5.1	Landfill	□ NA			
5.5.2	Land treatment/ application farming	□ NA			
5.5.3	Surface impoundment	□ NA			
5.5.4	Other disposal	□ NA			

□ Check here only if additional Section 5.3 information is provided on page 5 of this form.

EPA Form 9350-1 (Rev. 12/4/92) - Previous editions are obsolete.

Range Codes: A = 1 - 10 pounds; B = 11
C = 500 - 999 pounds.

⊕EPA
United States
Environmental Protection
Agency

EPA FORM R

PART II. CHEMICAL-SPECIFIC INFORMATION (CONTINUED)

TRI FACILITY ID NUMBER

Toxic Chemical, Category, or Generic Name

SECTION 5.3   ADDITIONAL INFORMATION ON RELEASES OF THE TOXIC CHEMICAL TO THE ENVIRONMENT ON-SITE

5.3	Discharges to receiving streams or water bodies (enter one name per box)	A. Total Release (pounds/year) (enter range code from instructions or estimate)	B. Basis of Estimate (enter code)	C. % From Stormwater
5.3.___	Stream or Water Body Name			
5.3.___	Stream or Water Body Name			
5.3.___	Stream or Water Body Name			

SECTION 6.   TRANSFERS OF THE TOXIC CHEMICAL IN WASTES TO OFF-SITE LOCATIONS

6.1   DISCHARGES TO PUBLICLY OWNED TREATMENT WORKS (POTW)

6.1.A   Total Quantity Transferred to POTWs and Basis of Estimate

6.1.A.1 Total Transfers (pounds/year) (enter range code or estimate)	6.1.A.2 Basis of Estimate (enter code)

6.1.B   POTW Name and Location Information

6.1.B.___ POTW Name	6.1.B.___ POTW Name
Street Address	Street Address
City / County	City / County
State / Zip Code	State / Zip Code

If additional pages of Part II, Sections 5.3 and/or 6.1 are attached, indicate the total number of pages in this box [ ] and indicate which Part II, Sections 5.3/6.1 page this is, here. [ ]

(example: 1, 2, 3, etc.)

EPA Form 9350-1 (Rev. 12/4/92) - Previous editions are obsolete.

Range Codes: A = 1 - 10 pounds; B = 11
C = 500 - 999 pounds.

EPA FORM R
PART II. CHEMICAL-SPECIFIC INFORMATION (CONTINUED)

EPA — United States Environmental Protection Agency

TRI FACILITY ID NUMBER
Toxic Chemical, Category, or Generic Name

SECTION 6.2   TRANSFERS TO OTHER OFF-SITE LOCATIONS

6.2.___ Off-site EPA Identification Number (RCRA ID No.)

Off-Site Location Name

Street Address

City   County

State   Zip Code   Is location under control of reporting facility or parent company? ☐ Yes ☐ No

A. Total Transfers (pounds/year) (enter range code or estimate)	B. Basis of Estimate (enter code)	C. Type of Waste Treatment/Disposal/ Recycling/Energy Recovery (enter code)
1.	1.	1. M
2.	2.	2. M
3.	3.	3. M
4.	4.	4. M

SECTION 6.2   TRANSFERS TO OTHER OFF-SITE LOCATIONS

6.2.___ Off-site EPA Identification Number (RCRA ID No.)

Off-Site Location Name

Street Address

City   County

State   Zip Code   Is location under control of reporting facility or parent company? ☐ Yes ☐ No

A. Total Transfers (pounds/year) (enter range code or estimate)	B. Basis of Estimate (enter code)	C. Type of Waste Treatment/Disposal/ Recycling/Energy Recovery (enter code)
1.	1.	1. M
2.	2.	2. M
3.	3.	3. M
4.	4.	4. M

If additional pages of Part II, Section 6.2 are attached, indicate the total number of pages in this box ☐ and indicate which Part II, Section 6.2 page this is, here. ☐ (example: 1, 2, 3, etc.)

EPA Form 9350-1 (Rev. 12/4/92) - Previous editions are obsolete.

Range Codes: A = 1 - 10 pounds; B = 11 - 499 pounds; C = 500 - 999 pounds.

⊕EPA

United States
Environmental Protection
Agency

EPA FORM R

PART II. CHEMICAL-SPECIFIC
INFORMATION (CONTINUED)

TRI FACILITY ID NUMBER

Toxic Chemical, Category, or Generic Name

SECTION 7A. ON-SITE WASTE TREATMENT METHODS AND EFFICIENCY

☐ **Not Applicable (NA) - Check here if <u>no</u> on-site waste treatment is applied to any waste stream containing the toxic chemical or chemical category.**

a. General Waste Stream (enter code)	b. Waste Treatment Method(s) Sequence [enter 3-character code(s)]	c. Range of Influent Concentration	d. Waste Treatment Efficiency Estimate	e. Based on Operating Data?
7A.1a	7A.1b 1 ☐ 2 ☐ 3 ☐ 4 ☐ 5 ☐ 6 ☐ 7 ☐ 8 ☐	7A.1c	7A.1d %	7A.1e Yes ☐ No ☐
7A.2a	7A.2b 1 ☐ 2 ☐ 3 ☐ 4 ☐ 5 ☐ 6 ☐ 7 ☐ 8 ☐	7A.2c	7A.2d %	7A.2e Yes ☐ No ☐
7A.3a	7A.3b 1 ☐ 2 ☐ 3 ☐ 4 ☐ 5 ☐ 6 ☐ 7 ☐ 8 ☐	7A.3c	7A.3d %	7A.3e Yes ☐ No ☐
7A.4a	7A.4b 1 ☐ 2 ☐ 3 ☐ 4 ☐ 5 ☐ 6 ☐ 7 ☐ 8 ☐	7A.4c	7A.4d %	7A.4e Yes ☐ No ☐
7A.5a	7A.5b 1 ☐ 2 ☐ 3 ☐ 4 ☐ 5 ☐ 6 ☐ 7 ☐ 8 ☐	7A.5c	7A.5d %	7A.5e Yes ☐ No ☐

If additional copies of page 7 are attached, indicate the total number of pages in this box ☐ and indicate which page 7 this is, here. ☐ (example: 1, 2, 3, etc.)

EPA Form 9350-1 (Rev. 12/4/92) - Previous editions are obsolete.

EPA FORM R
PART II. CHEMICAL-SPECIFIC INFORMATION (CONTINUED)

United States
Environmental Protection
Agency

TRI FACILITY ID NUMBER

Toxic Chemical, Category, or Generic Name

SECTION 7B. ON-SITE ENERGY RECOVERY PROCESSES

☐ **Not Applicable (NA)** - Check here if <u>no</u> on-site energy recovery is applied to any waste stream containing the toxic chemical or chemical category.

Energy Recovery Methods [enter 3-character code(s)]

1 [ ]   2 [ ]   3 [ ]   4 [ ]

SECTION 7C. ON-SITE RECYCLING PROCESSES

☐ **Not Applicable (NA)** - Check here if <u>no</u> on-site recycling is applied to any waste stream containing the toxic chemical or chemical category.

Recycling Methods [enter 3-character code(s)]

1 [ ]   2 [ ]   3 [ ]   4 [ ]   5 [ ]

6 [ ]   7 [ ]   8 [ ]   9 [ ]   10 [ ]

EPA Form 9350-1 (Rev. 12/4/92) - Previous editions are obsolete.

⊕EPA
United States
Environmental Protection
Agency

EPA FORM R

PART II. CHEMICAL-SPECIFIC INFORMATION (CONTINUED)

TRI FACILITY ID NUMBER

Chemical, Category, or Generic Name

SECTION 8. SOURCE REDUCTION AND RECYCLING ACTIVITIES

All quantity estimates can be reported using up to two significant figures.	Column A 1991 (pounds/year)	Column B 1992 (pounds/year)	Column C 1993 (pounds/year)	Column D 1994 (pounds/year)
8.1 Quantity released *				
8.2 Quantity used for energy recovery on-site				
8.3 Quantity used for energy recovery off-site				
8.4 Quantity recycled on-site				
8.5 Quantity recycled off-site				
8.6 Quantity treated on-site				
8.7 Quantity treated off-site				
8.8 Quantity released to the environment as a result of remedial actions, catastrophic events, or one-time events not associated with production processes (pounds/year)				
8.9 Production ratio or activity index				

8.10 Did your facility engage in any source reduction activities for this chemical during the reporting year? If not, enter "NA" in Section 8.10.1 and answer Section 8.11.

	Source Reduction Activities [enter code(s)]	Methods to Identify Activity (enter codes)		
8.10.1		a.	b.	c.
8.10.2		a.	b.	c.
8.10.3		a.	b.	c.
8.10.4		a.	b.	c.

8.11 Is additional optional information on source reduction, recycling, or pollution control activities included with this report? (Check one box)   YES ☐   NO ☐

* Report releases pursuant to EPCRA Section 329(8) including "any spilling, leaking, pumping, pouring, emitting, emptying, discharging, injecting, escaping, leaching, dumping, or disposing into the environment." Do not include any quantity treated on-site or off-site.

EPA Form 9350 - 1 (Rev. 12/4/92) - Previous editions are obsolete.

Page _____ of _____ pages
Form Approved OMB No 2050—0072

Tier One

EMERGENCY AND HAZARDOUS CHEMICAL INVENTORY

Aggregate Information by Hazard Type

FOR OFFICIAL USE ONLY

ID #

Date Received

Important: Read instructions before completing form

Reporting Period   From January 1 to December 31, 19______

Facility Identification

Name _______________________________

Street _______________________________

City _____________ County _________ State _____ Zip _________

SIC Code [ ][ ][ ][ ]   Dun & Brad Number [ ][ ]-[ ][ ][ ][ ]-[ ][ ][ ][ ]

Emergency Contacts

Name _______________________________
Title _______________________________
Phone (____) _______________
24 Hour Phone (____) _______________

Name _______________________________
Title _______________________________
Phone (____) _______________
24 Hour Phone (____) _______________

[ ] Check if information below is identical to the information submitted last year.

Owner/Operator

Name _______________________________

Mail Address _______________________________

Phone (____) _______________

[ ] Check if site plan is attached

Physical Hazards

Hazard Type	Max Amount*	Average Daily Amount*	Number of Days On–Site	General Location
Fire	[ ][ ]	[ ][ ]	[ ][ ][ ]	
Sudden Release of Pressure	[ ][ ]	[ ][ ]	[ ][ ][ ]	
Reactivity	[ ][ ]	[ ][ ]	[ ][ ][ ]	

Health Hazards

Hazard Type	Max Amount*	Average Daily Amount*	Number of Days On–Site	General Location
Immediate (acute)	[ ][ ]	[ ][ ]	[ ][ ][ ]	
Delayed (Chronic)	[ ][ ]	[ ][ ]	[ ][ ][ ]	

Certification   *{Read and sign after completing all sections}*

I certify under penalty of law that I have personally examined and am familiar with the information submitted in pages one through _______, and that based on my inquiry of those individuals responsible for obtaining the information, I believe that the submitted information is true, accurate and complete.

____

Name and official title of owner/operator OR owner/operator's authorized representative

_______________   _______________
Signature   Date signed

*   Reporting Ranges

Range Code	Weight Range in Pounds From...	To...
01	0	99
02	100	999
03	1000	9,999
04	10,000	99,999
05	100,000	999,999
06	1,000,000	9,999,999
07	10,000,000	49,999,999
08	50,000,000	99,999,999
09	100,000,000	499,999,999
10	500,000,000	999,999,999
11	1 billion	higher than 1 billion

Revised June 1990

Page _______ of _______ pages
Form Approved OMB No 2050-0072

Tier Two

EMERGENCY AND HAZARDOUS CHEMICAL INVENTORY

Specific Information by Chemical

Facility Identification

Name _______________________
Street _______________________
City _____________ County _____________ State _______ Zip _______

SIC Code ☐☐☐☐   Dun & Brad Number ☐☐-☐☐☐☐-☐☐☐☐

FOR OFFICIAL USE ONLY
ID # _______________________
Date Received _______________________

Owner/Operator Name

Name _______________________ Phone (____) __________
Mail Address _______________________

Emergency Contact

Name _______________________ Title __________
Phone (____) __________ 24 Hr. Phone (____) __________

Name _______________________ Title __________
Phone (____) __________ 24 Hr. Phone (____) __________

Important: Read all instructions before completing form

Reporting Period From January 1 to December 31, 19 _____

☐ Check if information below is identical to the information submitted last year.

Chemical Description	Physical and Health Hazards (check all that apply)	Inventory	Container Type / Temperature / Pressure	Storage Codes and Locations (Non-Confidential) — *Storage Locations*	Optional
CAS ☐☐☐☐☐☐☐ ☐☐ ☐ Trade Secret ☐ Chem. Name _______________ Check all that apply: ☐ Pure ☐ Mix ☐ Solid ☐ Liquid ☐ Gas ☐ EHS EHS Name _______________	☐ Fire ☐ Sudden Release of Pressure ☐ Reactivity ☐ Immediate (acute) ☐ Delayed (chronic)	☐☐ Max. Daily Amount (code) ☐☐ Avg. Daily Amount (code) ☐☐ No. of Days On-site (days)			☐
CAS ☐☐☐☐☐☐☐ ☐☐ ☐ Trade Secret ☐ Chem. Name _______________ Check all that apply: ☐ Pure ☐ Mix ☐ Solid ☐ Liquid ☐ Gas ☐ EHS EHS Name _______________	☐ Fire ☐ Sudden Release of Pressure ☐ Reactivity ☐ Immediate (acute) ☐ Delayed (chronic)	☐☐ Max. Daily Amount (code) ☐☐ Avg. Daily Amount (code) ☐☐ No. of Days On-site (days)			☐
CAS ☐☐☐☐☐☐☐ ☐☐ ☐ Trade Secret ☐ Chem. Name _______________ Check all that apply: ☐ Pure ☐ Mix ☐ Solid ☐ Liquid ☐ Gas ☐ EHS EHS Name _______________	☐ Fire ☐ Sudden Release of Pressure ☐ Reactivity ☐ Immediate (acute) ☐ Delayed (chronic)	☐☐ Max. Daily Amount (code) ☐☐ Avg. Daily Amount (code) ☐☐☐ No. of Days On-site (days)			☐

Certification   *(Read and sign after completing all sections)*

I certify under penalty of law that I have personally examined and am familiar with the information submitted in pages one through _______, and that based on my inquiry of those individuals responsible for obtaining the information, I believe that the submitted information is true, accurate, and complete.

Name and official title of owner/operator OR owner/operator's authorized representative _______________ Signature _______________ Date signed _______________

Optional Attachments

☐ I have attached a site plan
☐ I have attached a list of site coordinate abbreviations
☐ I have attached a description of dikes and other safeguard measures

Tier Two

EMERGENCY AND HAZARDOUS CHEMICAL INVENTORY

Specific Information by Chemical

Facility Identification

Name ______________________________
Street ______________________________
City __________ County __________ State __________ Zip __________

SIC Code ☐☐☐☐   Dun & Brad Number ☐☐-☐☐☐-☐☐☐☐

FOR OFFICIAL USE ONLY
ID # ______________________________
Date Received ______________________________

Owner/Operator Name

Name __________________________ Phone ( )
Mail Address ______________________________

Emergency Contact

Name __________________________ Title __________
Phone ( ) __________ 24 Hr. Phone ( )

Name __________________________ Title __________
Phone ( ) __________ 24 Hr. Phone ( )

☐ Check if information below is identical to the information submitted last year.

Important: Read all instructions before completing form

Reporting Period   From January 1 to December 31, 19 _____

Confidential Location Information Sheet

Storage Codes and Locations
(Confidential)

Container Type | Temperature | Pressure | *Storage Locations* | Optional

CAS # ☐☐☐☐☐☐ ☐☐ ☐   Chem. Name

CAS # ☐☐☐☐☐☐ ☐☐ ☐   Chem. Name

CAS # ☐☐☐☐☐☐ ☐☐ ☐   Chem. Name

Certification   *(Read and sign after completing all sections)*

I certify under penalty of law that I have personally examined and am familiar with the information submitted in pages one through _________, and that based on my inquiry of those individuals responsible for obtaining the information, I believe that the submitted information is true, accurate, and complete.

Name and official title of owner/operator OR owner/operator's authorized representative _______________

Signature _______________   Date signed _______________

Optional Attachments

☐ I have attached a site plan
☐ I have attached a list of site coordinate abbreviations
☐ I have attached a description of dikes and other safeguard measures

BEFORE COPYING FORM, ATTACH SITE IDENTIFICATION LABEL OR ENTER:

SITE NAME ___________________________________

____

EPA ID NO. | | | | | | | | | | | | |

U.S. ENVIRONMENTAL PROTECTION AGENCY

1991 Hazardous Waste Report

FORM

IC

IDENTIFICATION AND CERTIFICATION

INSTRUCTIONS: Read the detailed instructions beginning on page 6 of the 1991 Hazardous Waste Report booklet before completing this form.

SEC. I Site name and location address. Complete items A through H. Check the box ☒ in items A, C, E, F, G, and H if same as label; if different, enter corrections. If label is absent, enter information. Instruction page 6

A. EPA ID No.
Same as label ☐ or ——➤ | | | | | | | | | | | | |

B. County

C. Site/company name
Same as label ☐ or ——➤

D. Has the site name associated with this EPA ID changed since 1989? ☐ 1 Yes ☐ 2 No

E. Street name and number. If not applicable, enter industrial park, building name or other physical location description.
Same as label ☐ or ——

F. City, town, village, etc.
Same as label ☐ or ——➤

G. State
Same as label ☐
| | |

H. Zip Code
Same as label ☐
| | | | | — | | | |

SEC. II Mailing address of site. Instruction page 6

A. Is the mailing address the same as the location address? ☐ 1 Yes (SKIP TO SEC. III) ☐ 2 No (GO TO BOX B)

B. Number and street name of mailing address

C. City, town, village, etc.

D. State
| | |

E. Zip Code
| | | | | — | | | |

SEC. III Name, title, and telephone number of the person who should be contacted if questions arise regarding this report. Instruction page 6

A. Please print: Last name   First name   M.I.

B. Title

C. Telephone
| | | | | | | | | — | | | |
Extension | | | | |

SEC. IV Enter the Standard Industrial Classification (SIC) Code that describes the principal products, group of products, produced or distributed, or the services rendered at the site's physical location. Enter more than one SIC Code only if no one industry description includes the combined activities of the site. Instruction page 7

A. | | | | |

B. | | | | |

C. | | | | |

D. | | | | |

SEC. V "I certify under penalty of law that this document and all attachments were prepared under my direction or supervision in accordance with a system designed to assure that qualified personnel properly gather and evaluate the information submitted. Based on my inquiry of the person or persons who manage the system, or those persons directly responsible for gathering the information, the information submitted is, to the best of my knowledge and belief, true, accurate and complete. I am aware that there are significant penalties under Section 3008 of the Resource Conservation and Recovery Act for submitting false information, including the possibility of fine and imprisonment for knowing violations."

A. Please print: Last name   First name   M.I.

B. Title

C. Signature

D. Date of signature
| | | | | | | | |
MO.   DAY   YR.

Page 1 of ________

EPA Form 8700--13A/B   (Revised 8-91)

OVER -->

Sec. VI - Generator Status

EPA ID NO. ☐☐☐☐☐☐☐☐☐☐☐☐

A. 1991 RCRA generator status
Instruction page 7
(CHECK ONE BOX BELOW)

- ☐ 1 LQG (SKIP TO SEC. VII)
- ☐ 2 SQG (SKIP TO SEC. VII)
- ☐ 3 CESQG (SKIP TO SEC. VII)
- ☐ 4 Non generator (CONTINUE TO BOX B)

B. Reason for not generating
Page 9
(CHECK ALL THAT APPLY)

- ☐ 1 Never generated
- ☐ 2 Out of business
- ☐ 3 Only excluded or delisted waste
- ☐ 4 Only non-hazardous waste
- ☐ 5 Periodic or occasional generator
- ☐ 6 Waste minimization activity
- ☐ 7 Other (SPECIFY COMMENTS IN BOX BELOW)

Sec. VII - On-Site Waste Management Status

A. RCRA permitted or interim status storage
Instruction page 10
☐

B. RCRA permitted or interim status treatment, disposal, or recycling
Page 10
☐

C. RCRA-exempt treatment, disposal, or recycling
Page 11
☐

Sec. VIII - Waste Minimization Activity during 1990 or 1991

A. Did this site begin or expand a source reduction activity during 1990 or 1991?
Instruction page 11

- ☐ 1 Yes
- ☐ 2 No

B. Did this site begin or expand a recycling activity during 1990 or 1991?
Page 12

- ☐ 1 Yes
- ☐ 2 No

C. Did this site systematically investigate opportunities for source reduction or recycling during 1990 or 1991?
Page 12

- ☐ 1 Yes
- ☐ 2 No

D. Did any of the factors listed below delay or limit this site's ability to initiate new or additional source reduction activities in 1990 or 1991?
Page 12
(CHECK YES OR NO FOR EACH ITEM)

Yes	No		
☐1	☐ 2	a.	Insufficient capital to install new source reduction equipment or implement new source reduction practices
☐1	☐ 2	b.	Lack of technical information on source reduction techniques applicable to the specific production processes
☐1	☐ 2	c.	Source reduction is not economically feasible: cost savings in waste management or production will not recover the capital investment
☐1	☐ 2	d.	Concern that product quality may decline as a result of source reduction
☐1	☐ 2	e.	Technical limitations of the production processes
☐1	☐ 2	f.	Permitting burdens
☐1	☐ 2	g.	Source reduction previously implemented – additional reduction does not appear to be technically feasible
☐1	☐ 2	h.	Source reduction previously implemented - additional reduction does not appear to be economically feasible
☐1	☐ 2	i.	Source reduction previously implemented - additional reduction does not appear to be feasible due to permitting requirements
☐1	☐ 2	j.	Other (SPECIFY COMMENTS IN BOX BELOW)

E. Did any of the factors listed below delay or limit this site's ability to initiate new or additional on-site or off-site recycling activities during 1990 or 1991?
Page 12
(CHECK YES OR NO FOR EACH ITEM)

Yes	No		
☐1	☐ 2	a.	Insufficient capital to install new recycling equipment or implement new recycling practice
☐1	☐ 2	b.	Lack of technical information on recycling techniques applicable to this site's specific production processes
☐1	☐ 2	c.	Recycling is not economically feasible: cost savings in waste management or production will not recover the capital investment
☐1	☐ 2	d.	Concern that product quality may decline as a result of recycling
☐1	☐ 2	e.	Requirements to manifest wastes inhibit shipments off site for recycling
☐1	☐ 2	f.	Financial liability provisions inhibit shipments off site for recycling
☐1	☐ 2	g.	Technical limitations of production processes inhibit shipments off site for recycling
☐1	☐ 2	h.	Technical limitations of production processes inhibit on-site recycling
☐1	☐ 2	i.	Permitting burdens inhibit recycling
☐1	☐ 2	j.	Lack of permitted off-site recycling facilities
☐1	☐ 2	k.	Unable to identify a market for recyclable materials
☐1	☐ 2	l.	Recycling previously implemented – additional recycling does not appear to be technically feasible
☐1	☐ 2	m.	Recycling previously implemented – additional recycling does not appear to be economically feasible
☐1	☐ 2	n.	Recycling previously implemented - additional recycling does not appear to be feasible due to permitting requirements
☐1	☐ 2	o.	Other (SPECIFY COMMENTS IN BOX BELOW)

Comments:

Page 2 of _____

BEFORE COPYING FORM, ATTACH SITE IDENTIFICATION LABEL OR ENTER:

SITE NAME _______________________________

____

EPA ID NO. |__|__|__| |__|__|__| |__|__|__| |__|__|

U.S. ENVIRONMENTAL PROTECTION AGENCY

1991 Hazardous Waste Report

FORM GM

WASTE GENERATION AND MANAGEMENT

INSTRUCTIONS: Read the detailed instructions beginning on page 13 of the 1991 Hazardous Waste Report booklet before completing this form.

Sec. I

A. Waste description
Instruction Page 15

B. EPA hazardous waste code
Page 15
|__|__|__|__| |__|__|__|__|
|__|__|__|__| |__|__|__|__| |__|__|__|__|

C. State hazardous waste code
Page 15
|__|__|__|__|__| |__|__|__|__|__|

D. SIC code
Page 16
|__|__|__|__|

E. Origin code |__|
Page 16
System type |M|__|__|__|

F. Source code
Page 17
|A|__|__|

G. Point of measurement
Page 17
|__|__|

H. Form code
Page 17
|B|__|__|__|

I. RCRA-radioactive mixed
Page 17
|__|__|

J. Reported TRI constituent
Page 18
|__|__|

K. CAS numbers
Page 18
1. |__|__|__|__|__|__|-|__|__|-|__| 2. |__|__|__|__|__|__|-|__|__|-|__|
3. |__|__|__|__|__|__|-|__|__|-|__| 4. |__|__|__|__|__|__|-|__|__|-|__| 5. |__|__|__|__|__|__|-|__|__|-|__|

Sec. II

A. Quantity generated in 1990
instruction Page 18
|__|__|__|__|__|__|__|__|__|.|__|

B. Quantity generated in 1991
Page 18
|__|__|__|__|__|__|__|__|__|.|__|

C. UOM   Density
Page 19
|__|__| |__|__|.|__|__|__|
☐ 1 lbs/gal ☐ 2 sg

D. Did this site do any of the following to this waste: treat on site, dispose on site, recycle on site, or discharge to a sewer/POTW?
Page 19
☐ 1 Yes (CONTINUE TO SYSTEM 1)
☐ 2 No   (SKIP TO SEC. III)

ON-SITE SYSTEM 1

On-site system type
Page 19
|M|__|__|__|

Quantity treated, disposed or recycled on site in 1991
|__|__|__|__|__|__|__|__|.|__|

ON-SITE SYSTEM 2

On-site system type
Page 19
|M|__|__|__|

Quantity treated, disposed or recycled on site in 1991
|__|__|__|__|__|__|__|__|.|__|

Sec. III

A. Was any of this waste shipped off site in 1991?
Instruction Page 20
☐ 1 Yes (CONTINUE TO BOX B)
☐ 2 No (SKIP TO SEC. IV)

Site 1

B. EPA ID No. of facility waste was shipped to
Page 20
|__|__|__|__|__|__|__|__|__|__|__|__|

C. System type shipped to
Page 20
|M|__|__|__|

D. Off-site availability code
Page 21
|__|__|

E. Total quantity shipped in 1991
Page 21
|__|__|__|__|__|__|__|__|__|.|__|

Site 2

B. EPA ID No. of facility waste was shipped to
Page 20
|__|__|__|__|__|__|__|__|__|__|__|__|

C. System type shipped to
Page 20
|M|__|__|__|

D. Off-site availability code
Page 21
|__|__|

E. Total quantity shipped in 1991
Page 21
|__|__|__|__|__|__|__|__|__|.|__|

Sec. IV

A. Did new activities in 1991 result in minimization of this waste?
Instruction Page 22
☐ 1 Yes (CONTINUE TO BOX B)
☐ 2 No   (THIS FORM IS COMPLETE)

B. Activity
Page 22
|W|__|__| |W|__|__|
|W|__|__| |W|__|__|

C. Other effects
Page 22
☐ 1 Yes
☐ 2 No

D. Quantity recycled in 1991 due to new activities
Page 23
|__|__|__|__|__|__|__|__|.|__|

E. Activity/production index
Page 23
|__|__|__|.|__|

F. 1991 Source reduction quantity
Page 24
|__|__|__|__|__|__|__|__|__|.|__|

Comments:

Page _____ of _____

Appendix E   Labor Contacts

Committees on Occupational Safety and Health (COSH) Contacts

Alaska

Alaska Health Project
1818 W. Northern Lights Blvd., Suite 103
Anchorage, AK 99517
(907) 276-2864

California

BACOSH/Worksafe (Bay Area)
8400 Enterprise Way, #104
Oakland, CA 94621
(415) 638-1174

LACOSH (Los Angeles)
600 South New Hampshire
Los Angeles, CA 90005
(213) 383-4416

SACOSH (Sacramento)
c/o Fire Fighters, Local 522
3101 Stockton Blvd.
Sacramento, CA 95820
(916) 924-8060 (Frank Damiata)

SCCOSH (Santa Clara)
760 N. 1st Street
San Jose, CA 95112
(408) 998-4050

Connecticut

ConnectiCOSH
P.O. Box 31107
Hartford, CT 06103
(203) 549-1877

District of Columbia

Alice Hamilton Occupational Health Center
410 Seventh Street, SE
Washington, DC 20003
(202) 543-0005

Illinois

CACOSH (Chicago Area)
37 South Ashland
Chicago, IL 60607
(312) 666-1611

Maine

Maine Labor Group on Health
Box V
Augusta, ME 04330
(207) 622-7823

Massachusetts

MassCOSH
555 Armory Street
Boston, MA 02130
(617) 524-6686

Western MassCOSH
458 Bridge Street
Springfield, MA 01103
(413) 731-0760

Michigan

SEMCOSH (Southeast Michigan)
2727 Second Street
Detroit, MI 48206
(313) 961-3345

New Hampshire
NHCOSH c/o NH AFL-CIO
110 Sheep Davis Road
Pembroke, NH 03275
(603) 228-3711

New York
ALCOSH (Allegheny)
100 East Second Street
Jamestown, NY 14701
(716) 488-0720

CYNCOSH (Central NY)
615 W. Genesee Street
Syracuse, NY 13204
(315) 471-6187

ENYCOSH (Eastern NY)
c/o Larry Rafferty
121 Erie Blvd.
Schenectady, NY 12305
(518) 374-4308

NYCOSH (New York)
275 Seventh Avenue, 8th Floor
New York, NY 10001
(212) 627-3900
(914) 939-5612 (Lower Hudson)
(516) 755-2400 (Long Island)

ROCOSH (Rochester)
797 Elmwood Avenue, #4
Rochester, NY 14620
(716) 244-0420

WNYCOSH (Western NY)
450 Grider Street
Buffalo, NY 14215
(716) 897-2110

North Carolina
NCOSH (North Carolina)
P.O. Box 2514
Durham, NC 27715
(919) 286-9249

Oregon (forming)
c/o Dick Edgington
ICWU-Portland
7440 SW 87 Street
Portland, OR 07223

Pennsylvania
PhilaPOSH (Philadelphia Project OSH)
3001 Walnut Street, 5th Floor
Philadelphia, PA 19104
215-386-7000

Rhode Island
RICOSH (Rhode Island)
741 Westminster Street
Providence, RI 02903
(401) 751-2015

Tennessee
TNCOSH (Tennessee)
1515 East Magnolia, Suite 406
Knoxville, TN 37917
(615) 525-3147

Texas
TexCOSH
c/o Karyl Dunson
5735 Regina
Beaumont, TX 77706
(409) 898-1427

Vermont (forming)
c/o Kathy Stanford
23 Highland Avenue, #5
Barre, VT 05641
(802) 476-9484

Washington (forming)
c/o Lin Nelson
833B Steele Street
Olympia, WA 98506
(206) 956-1358

Wisconsin
WisCOSH (Wisconsin)
734 North 26th Street
Milwaukee, WI 53230
(414) 643-0928

Canada – Ontario
WOSH (Windsor OSH)
1731 Wyandotte Street
East Windsor, Ontario N8Y 1C9
(519) 254-4192

COSH-Related Groups

California
Labor Occupational Health Program
Institute of Industrial Relations
University of California
Berkeley, CA 94720
(415) 642-5507

District of Columbia
Workers Institute for Occupational Safety
and Health
1126 16th Street, NW, Room 403
Washington, DC 20036
(202) 887-1980

Louisiana
Labor Studies Program/LA Watch Institute of
Human Relations
Loyola University, Box 12
New Orleans, LA 70118
(504) 861-5830

New Jersey
New Jersey Work Environment Council
452 East Third Street
Moorestown, NJ 08057
(609) 866-9405

Ohio
Greater Cincinnati Occupational Health Center
10475 Reading Road
Cincinnati, OH 45241
(513) 541-0561

West Virginia
Institute of Labor Studies
710 Knapp Hall
West Virginia University
Morgantown, WV 26506
(304) 293-3323

U.S. Department of Labor, Occupational Safety and Health Administration Regional Contacts

Region I (CT, MA, ME, NH, RI, VT*)
John Miles, Regional Administrator
133 Portland Street, 1st Floor
Boston, MA 02114
(617) 565-7164

Region II (NJ, NY,* Puerto Rico,* Virgin Islands*)
James W. Stanley, Regional Administrator
201 Varick Street, Room 670
New York, NY 10014
(212) 337-2378

Region III (DC, DE, MD,* PA, VA,* WV)
Linda R. Anku, Regional Administrator
Gateway Building, Suite 2100
3535 Market Street
Philadelphia, PA 19104
(215) 596-1201

Region IV (AL, FL, GA, KY,* MS, NC,* SC,* TN*)
R. Davis Layne, Regional Administrator
1375 Peachtree Street, NE, Suite 587
Atlanta, GA, 30367
(404) 347-3573

Region V (IL, IN,* MI,* MN,* OH, WI)
Michael G. Connors, Regional Administrator
230 South Dearborn Street, Room 3244
Chicago, IL 60604
(312) 353-2220

Region VI (AZ, LA, NM,* OK, TX)
Gilbert J. Saulter, Regional Administrator
525 Griffin Street, Room 602
Dallas, TX 75202
(214) 767-4791

Region VII (IA,* KS, MO, NE)
John T. Phillips, Regional Administrator
911 Walnut Street, Room 406
Kansas City, MO 64106
(816) 426-5861

Region VIII (CO, MT, ND, SD, UT,* WY*)
Byron R. Chadwick, Regional Administrator
Federal Building, Room 1576
1961 Stout Street
Denver, CO 80294
(303) 844-3061

Region IX (American Samoa, AZ,* CA,* Guam, HI,* NV,* Trust Territories of the Pacific)
Frank Strasheim, Regional Administrator
71 Stevenson Street, Room 415
San Francisco, CA 94105
(415) 744-6670

Region X (AK,* ID, OR,* WA*)
James W. Lake, Regional Administrator
1111 Third Avenue, Suite 715
Seattle, WA 98101-3212
(206) 553-5930

* These states and territories operate their own OSHA-approved
job safety and health programs (Connecticut and New York plans
cover public employees only). States with approved programs
must have a standard that is identical to, or at least as effective as,
the federal standard.

Appendix F — Using the Freedom of Information Act

As you have seen throughout this guide, it is often necessary to make a Freedom of Information Act (FOIA) request to obtain publicly available information. FOIA generally establishes the public's right to obtain information from Federal government agencies. In FOIA, Congress created a specific procedure through which any person may exercise the public's general right to request and obtain access to particular agency records: Procedures vary among states and at the federal level. Nonetheless, the pointers below should apply to any FOIA request you make:

- Always verify with the state or Federal official that you have the correct contact in making your FOIA request.
- Write legibly or type the letter, and include the date. Always make copies of the FOIA letters.
- State that you are making an information request under Section 552 of the Freedom of Information Act.
- Be specific about the information you want. (Use the worksheets in this guide to help you list the data you are requesting).
- Ask to be notified of the size and cost of the order before it is filled by the state or federal official.
- Ask if you are eligible for a fee waiver.
- If you must follow up on your written request, keep notes of any conversations with agency officials.

It is highly recommended that you obtain a copy of the following manual before making any FOIA request. It contains many of the pointers above, plus an in-depth description of the FOIA process, including sample FOIA letters: Center for National Security Studies. *Using the Freedom of Information Act: A Step-by-Step Guide.* (Center for National Security Studies, 122 Maryland Avenue, NE, Washington, DC 20002. $4.)

National and Regional EPA Freedom of Information Contacts

The following Environmental Protection Agency contacts may be used for your FOIA requests. Call the office in your region to find out if there is a specific federal office or local contact where your FOIA request should be directed.

US Environmental Protection Agency
Freedom of Information Act Office
A-101
401 M Street, SW
Washington, DC 20460
Attention: Jeralene Greene

Region 1 (CT, MA, ME, NH, RI, VT)
US EPA Region 1
Freedom of Information Act Office
JFK Federal Building (RPA-74)
Boston, MA 02203
(617) 565-3187

Region 2 (NJ, NY, Puerto Rico, US Virgin Islands)
US EPA Region 2
Freedom of Information Act Office
Office of External Programs
26 Federal Plaza, Room 905
New York, NY 10278
(212) 264-2515

Region 3 (DE, DC, MD, PA, VA, WV)
US EPA Region 3
Freedom of Information Act Office (3PA00)
841 Chestnut Street
Philadelphia, PA 19107
(215) 597-0798

Region 4 (AL, FL, GA, KY, MS, NC, SC, TN)
US EPA Region 4
Freedom of Information Act Office
345 Courtland Street, NE
Atlanta, GA 30365
(404) 347-4332

Region 5 (IL, IN, MI, MN, OH, WI)
US EPA Region 5
Freedom of Information Act Office
77 West Jackson Street
Chicago, IL 60604
(312) 886-7067

Region 6 (AR, LA, NM, OK, TX)
US EPA Region 6
Freedom of Information Act Office
1455 Ross Avenue (6M-ASC)
Dallas, TX 75202-7233
(214) 655-6558

Region 7 (IA, KS, MO, NE)
US EPA Region 7
Freedom of Information Act Office
726 Minnesota Avenue
Kansas City, KS 66101
(913) 551-7586

Region 8 (CO, MT, ND, SD, UT, WY)
US EPA Region 8
Freedom of Information Act Office (80EA)
999 18th Street, Suite 500
Denver, CO 80202-2405
(303) 294-1116

**Region 9 (AZ, CA, HI, NV, American Samoa,
     Guam, Commonwealth of the North Mariana
     Islands, Trust Territory of the Pacific Islands)**
US EPA Region 9
Freedom of Information Act Office
75 Hawthorne Street
San Francisco, CA 94105
(415) 744-1581

Region 10 (AK, ID, OR, WA)
US EPA Region 10
Freedom of Information Act Office
   (MD-103)
1200 Sixth Avenue
Seattle, WA 98101
(206) 442-4280

Appendix G Glossary

Batch processing. Production of a single product at different times, or production of different products at different times, using the same equipment, to meet changing customer needs. (See continuous processing.)

CAS number. The American Chemical Society's Chemical Abstract Services unique number for each chemical; a given chemical may have more than one name but only one CAS number.

Chemical Inventory Forms (Emergency Hazardous Chemical Inventory Forms). Forms containing the daily and annual quantities of hazardous substances, and location of these substances at the facility. Chemical inventory forms are submitted to the US Environmental Protection Agency, Local Emergency Planning Committees (LEPCs), and local fire departments under the Emergency Planning and Community Right-to-Know Act (EPCRA).

Chemical substitution. Replacement of toxic or hazardous chemicals with nonhazardous, nontoxic, or less hazardous or less toxic ones in both production and nonproduction processes. (Also called material substitution or raw material substitution.)

Clean Air Act. As amended in 1991, provides the basic legal authority for US air pollution control programs, and is designed to enhance the quality of air resources.

Comprehensive Environmental Response, Compensation and Liability Act (CERCLA). As amended by the Superfund Amendments and Reauthorization Act of 1986 (SARA), establishes a program to deal with the release of hazardous substances in spills and from inactive or abandoned disposal sites.

Continuous processing. Production in which process equipment is dedicated to continuous production of a single product. Continuous process operations usually generate less waste per pound of product than do batch process operations. (See batch processing.)

Cost accounting. An account of all costs associated with the generation of a waste stream at the point at which it is generated. The account is done on a multimedia, chemical-specific basis and allows plant management to identify the contributions of each individual process to the plant's total waste generation. The following specific costs may be allocated to the individual process: materials costs (i.e., the costs of starting materials and wasted products lost); environmental handling costs (e.g., capital and operational expenses for treatment facilities, regulatory and compliance costs, waste transportation and disposal costs); insurance costs and future liabilities from hazardous wastes (e.g., from accidents, worker illness, or waste site cleanups). In addition to these specific costs, companies may also consider less quantifiable but important costs, such as those involved with public and customer relations related to waste problems.

Criteria Pollutants. Six common air pollutants (sulfur dioxide, carbon monoxide, particulates, nitrogen oxides, ozone, and lead) for which EPA set national ambient air quality standards (NAAQS) before 1990 under Section 108 of the Clean Air Act.

Deep well injection. Disposal of hazardous waste deep underground, presumably below groundwater supplies; considered by many to be a very dangerous and undesirable method of waste management. (Also called underground injection.)

Disposal. See hazardous waste disposal.

Discharge Monitoring Report (DMR). A periodic monitoring report of water discharges at a plant that must be submitted to state agencies pursuant to the Federal Water Pollution Control Act.

Emergency Planning and Community Right-to-Know Act (EPCRA), Title III of the Superfund Amendments and Reauthorization Act (SARA). Enacted by Congress in 1986, this law gave the public significant new rights to find out about the toxic or hazardous chemicals stored, used, and released throughout the country. In particular, Section 313 of Title III created the Toxics Release Inventory (TRI) which provides public data on "routine" chemical releases from industries across the United States.

End-of-the-pipe. The point, at the end of the production process, at which all products and waste products have been made, and the waste products are being released (through a pipe, smokestack, or other release point); usually used as an adjective to refer to a pollution control strategy.

Energy recovery. Obtaining energy from the controlled incineration of solid waste.

Equipment changes. Modifications of and additions to equipment used in any stage of the manufacturing process (e.g., equipment used for storing, moving, mixing, or reacting chemicals) in order to reduce the amount of waste generated. (Also called technology/equipment modification.)

Extremely hazardous substance. Chemical substance on the "Extremely Hazardous Substance List" created under Section 304 of the Emergency Planning and Community Right-to-Know Act (EPCRA). The list contains 360 chemicals that could cause serious human health effects from short-term exposures such as accidental air releases.

Facility-level materials accounting. A comparison of the inputs of individual chemicals (the amount existing in plant inventory plus the amount entering the plant) with the outputs (the amount consumed in plant processes plus the amount shipped out in products) to determine the amounts released to the environment.

Form R. Form on which companies report Toxics Release Inventory data to state and federal environmental officials under the Emergency Planning and Community Right-to-Know Act.

Fugitive air emissions. Air pollutants released through leaky valves, evaporation from tanks, and other unintentional release points.

Hazardous substance. (See Box 1-2, Chapter 1.)

Hazardous waste. Defined in the Resource Conservation and Recovery Act (RCRA) as a solid waste, or combination of solid wastes, that, because of quantity, concentration, or physical, chemical, or infectious characteristics, may cause or significantly contribute to an increase in mortality or an increase in serious irreversible, or incapacitating reversible, illness, or pose a substantial present or potential hazard to human health or the environment when improperly treated, stored, transported, disposed of, or otherwise managed.

Hazardous waste disposal. The incineration, long-term storage, treatment, discharge, deposit, injection, dumping, spilling, leaking, or placing of a hazardous waste into or on land or water.

Hazardous waste treatment. Any method, technique, or process that changes the physical, chemical, or biological composition of any hazardous waste and so renders it non-hazardous, safer for transport, capable of recovery and/or storage, or reduces its volume.

In-process recycling. (Also called closed-loop recycling.) Recycling occurring as an integral part of the production process. For example, moving a waste stream from one end of an operation to a point near the front end in a closed-loop system for reuse as a raw material in that same operation.

Land treatment. Application or incorporation of hazardous waste into the soil surface.

LEPC. See Local Emergency Planning Committee.

Local Emergency Planning Committee (LEPC). The federal Emergency Planning and Community Right-to-Know (EPCRA) law required the creation of local committees, in areas designated by the State Emergency Response Commission (SERC), comprised of representatives from the following groups: elected officials, police officers, firefighters, health organizations, hospitals, transportation officials, the media, community groups, and industrial facilities. LEPCs are required to gather the following information and make it available to the public: material safety data sheets, chemical inventories, community emergency plans, and accidental releases of extremely hazardous substances.

Materials accounting. A systematic tracking of raw materials and products as they move sequentially from one end of the plant to the other; not as quantitatively rigorous as a materials balance.

Materials balance. Quantitative assessment of chemical inputs and outputs of individual processes that aims to account for every pound of a chemical (a) shipped to the process, (b) created or destroyed in the process, (c) delivered as a product from the process, and (d) wasted (regardless of whether it is an air, water, or solid waste); if the amount of waste identified does not equal the difference between the amount of the chemical entering (or being created in) and leaving (or being consumed in) the process, then other sources of waste must exist and should be identified.

Material Safety Data Sheet (MSDS). Part of the Hazard Communication Standards (HCS) set up by the US Occupational Safety and Health Administration (OSHA) to protect workers from chemical hazards. The MSDS provides the chemical composition of the substance used, its trade name, name of the manufacturer, hazards associated with the substance, and precautions that workers should take to avoid such hazards. MSDSs are prepared by the manufacturer of the chemical or product. They are kept on site at the plant and made available to workers and the public by the employer.

MSDS. See Material Safety Data Sheet.

Multimedia. Applying to all environmental media: land, water, and air.

National Pollutant Discharge Elimination System (NPDES). The national program established under the Federal Water Pollution Control Act, which requires all point source discharges into any body of water to be permitted by EPA or the designated state agency. Minimum pretreatment requirements for such discharges are established under the program.

Non-product (hazardous) outputs. All hazardous substances or hazardous wastes generated prior to storage, recycling, treatment, control, or disposal that are not intended for use as a product.

Off-site Transfers. The shipment of hazardous waste to facilities where it is treated, stored, or disposed of.

Occupational Safety and Health Act. Enacted

in 1970 to improve working conditions in the nation.

Occupational Safety and Health Administration (OSHA). The federal agency responsible for the implementation and enforcement of the Occupational Safety and Health Act.

Operational changes. Changes in the way hazardous materials are handled at a plant (e.g., careful observation and control of materials, process conditions, and employee habits in order to minimize spills, process upsets, or the use of excessive amounts of chemicals) that can reduce the generation of waste.

Out-of-process recycling. See recycling.

PMN. See Premanufacture Notice.

POTW. See Publicly owned treatment works.

Point source emissions. Air pollutants released through smokestacks, vents, and other intentional release points.

Pollution Prevention Act of 1990. Establishes a national policy of hierarchical environmental protection, stating that pollution should be prevented or reduced at the source whenever feasible.

Premanufacture Notice (PMN). A form that must be completed by the manufacturer of a new chemical substance pursuant to the requirements of the Toxic Substances Control Act. The PMN contains chemical identity data (including how the chemical is used in the plant), general information about the manufacturer, and occupational exposure, environmental release, and disposal.

Priority Pollutants. A list of 126 water pollutants regulated by the Clean Water Act Amendment of 1977 as toxic chemicals that are particularly harmful to one or more forms of animal or plant life. They are primarily organics and metals. Organic pollutants include pesticides, solvents, polychlorinated biphenyls (PCBs), and dioxin. Metals include lead, silver, mercury, copper, chromium, zinc, nickel, and cadmium.

Process changes. Any change in the production process that reduces the generation of waste, ranging from simple alterations of process conditions such as temperature and pressure to discovery of new chemical pathways and production technologies.

Process-level source reduction inventory. A series of steps enabling company management to identify the sources and activities, on an individual process basis, that lead to toxic or hazardous substances or the generation of waste.

Product changes. Changes in the product itself that can be achieved without changing the fundamental manufacturing process and that reduce the generation of waste (e.g., creating a chemical product in the form of pellets rather than as a powder can reduce the generation of waste dusts as the material is transferred during final packaging operations). (Also called product reformulation.)

Production Ratio or Activity Index. A method for measuring the quantity of waste generated in relation to the quantity of product produced.

Publicly owned treatment works (POTW). Public sewage treatment facilities.

RCRA. See Resource Conservation and Recovery Act.

Recycling. The reuse of by-products, or reclamation of components of by-products, that might otherwise be disposed of in the environment. As currently used by the public, recycling includes everything from the creation of new paper from old paper to the burning of used oil and chemicals in furnaces.

Reportable Quantity (RQ). The minimum quantity of hazardous waste generated as a result of a discharge or spill that must be reported to a government agency.

Resource Conservation and Recovery Act (RCRA). The federal "cradle to grave" regulations for solid waste, both hazardous and non-hazardous (garbage). It includes certain wastewaters destined for land disposal.

Right-to-Know. A term usually referring to a series of laws, regulations, or databases that provide industry-related information to the public.

SARA. See Superfund Amendments and Reauthorization Act.

SIC (Standard Industrial Classification). The system the Federal government uses to classify US companies according to the products they produce (e.g., the chemical and allied products industry is assigned SIC code 28, with individual industries in this category assigned four-digit codes that begin with 28).

Source reduction. A strategy for reducing pollution that involves preventing the generation of waste in the first place rather than cleaning it up, treating it, or recycling after it has been produced.

State Emergency Response Commission (SERC). State bodies, required under the federal Emergency Planning and Community Right-to-Know Act, that coordinate the Local Emergency Planning Committees.

Superfund Amendments and Reauthorization Act (SARA). A 1986 federal law amending the original "Superfund" law. It is the primary mechanism for funding the clean-up of hazardous waste sites. Title III of this law is called the Emergency Planning and Community Right-to-Know Act (EPCRA). Section 313 of EPCRA contains the Toxics Release Inventory requirements.

Surface impoundment. A facility or part of a facility that is a natural topographic depression, man-made excavation, or diked area formed primarily of earthen materials (although it may be lined with synthetic materials), which is designed to hold an accumulation of liquid wastes or wastes containing free liquids, and which is not an injection well. Examples of surface impoundments are holding, storage, settling, and aeration pits, ponds, and lagoons.

Toxic substance. (See Box 1-2, Chapter 1.)

Toxics Release Inventory (TRI). The US Environmental Protection Agency's annual inventory of the amounts of about 320 chemicals released to the air, water, or land, or transferred off-site from the 20,000 or so largest manufacturing facilities using or manufacturing these chemicals in the United States. TRI provisions are found in Section 313 of the Emergency Planning and Community Right-to-Know Act, which is Title III of the 1986 Superfund Amendments and Reauthorization Act (SARA).

Toxic Substances Control Act (TSCA). A federal law enacted in 1976 to establish requirements for identifying and controlling potential toxic chemical hazards to human health and the environment. Programs developed under TSCA now gather information about the toxicity of certain chemicals, the extent of human and environmental exposure, and the potential risks of such exposure.

Toxics use reduction (TUR). Activities where the intent is to reduce, avoid, or eliminate the use of toxics in processes and/or products so as to reduce overall risks to the health of workers, consumers, and the environment without shifting risks among workers, consumers, or parts of the environment. TUR technologies coincide with source reduction technologies (i.e., product reformulation, raw-material substitution, and changes in operations, equipment, or process).

Toxic wastes. Specific chemicals or chemical mixtures that are thought to be injurious to human or environmental health, regardless of regulatory status or environmental medium into which they are released.

Treatment. See hazardous waste treatment

TRI. See Toxics Release Inventory.

Waste minimization. As defined in a 1986 Environmental Protection Agency report to Congress, "the reduction, to the extent feasible, of hazardous waste that is generated or subsequently treated, stored, or disposed of. It includes any source reduction or recycling activity undertaken by a waste generator that results in either (1) the reduction of total volume or quantity of hazardous waste, or (2) the reduction of toxicity of hazardous waste, or both, so long as such reduction is consistent with the goal of minimizing present and future threats to human health and the environment." Waste minimization is not a multimedia approach. It is concerned only with hazardous wastes as defined under the Resource Conservation and Recovery Act and does not include air emissions.

Water Pollution Control Act (Clean Water Act). As amended, provides the basic legal authority for federal water pollution control programs.

About the Authors

Marian Wise

Marian Wise is an attorney and senior associate with INFORM's Chemical Hazards Prevention Program. She has extensive experience in pollution prevention, federal and state right-to-know issues, and clean water litigation. Ms. Wise heads INFORM's pollution prevention outreach efforts in the Great Lakes states.

Before joining INFORM, Ms. Wise was an environmental attorney with the New Jersey Public Interest Research Group, responsible for research, public education, litigation, and advocacy in NJPIRG's toxics program. Most notably, she coordinated the group's grassroots campaign that led to passage of the New Jersey Pollution Prevention Act.

Ms. Wise was appointed by New Jersey Gov. James Florio to be environmental representative on both the Governor's Right-to-Know Advisory Council and the judging panel for the Governor's Award for Outstanding Achievement in Pollution Prevention. She has co-authored articles on toxics and pollution prevention, and reports analyzing the federal and state right-to-know data.

Marian Wise received her law degree from Rutgers School of Law and her undergraduate degree from New York University.

Lauren Kenworthy

Lauren Kenworthy is an environmental consultant and former legislative assistant to then-Rep. Howard E. Wolpe (D-MI). She drafted the Waste Reduction Act, which was introduced in Congress in 1986 by Rep. Wolpe. She was also responsible for educating congressional staff and members of the public about issues and options for promoting source reduction of industrial toxic wastes. Before working on Capitol Hill, Ms. Kenworthy was a research assistant at the Natural Resources Defense Council.

Ms. Kenworthy is working toward her Ph.D. at the University of Maryland. She received her bachelor's degree with honors from Yale University.

INFORM Publications and Membership

Publications

Chemical Hazards Prevention
Selected Publications

Environmental Dividends: Cutting More Chemical Wastes (Mark H. Dorfman, Warren R. Muir, Ph.D., and Catherine G. Miller, Ph.D.), 1992, 288 pp., $75.

Tackling Toxics in Everyday Products: A Directory of Organizations (Nancy Lilienthal, Michèle Ascione, Adam Flint), 1992, 180 pp., $19.95.

Cutting Chemical Wastes: What 29 Organic Chemical Plants Are Doing to Reduce Hazardous Wastes (David J. Sarokin, Warren Muir, Ph.D., Catherine G. Miller, Ph.D., and Sebastian R. Sperber), 1985, 548 pp., $47.

Toward a More Informed Public: Recommendations for Improving the Toxics Release Inventory (Jacqueline B. Courteau and Nancy Lilienthal), 1991, 22 pp., $10.

Pollution Prevention Through Technical Assistance: One State's Experience (Mark H. Dorfman and John Riggio), 1990, 72 pp., $15.

Promoting Hazardous Waste Reduction: Six Steps States Can Take (Warren R. Muir, Ph.D., and Joanna D. Underwood), 1987, 24 pp., $3.50.

Municipal Solid Waste
Selected Publications

Making Less Garbage: A Planning Guide for Communities (Bette K. Fishbein and Caroline Gelb), 1992, 180 pp., $30.

Business Recycling Manual (copublished with Recourse Systems, Inc.), 1991, 202 pp., $85.

Burning Garbage in the US: Practice vs. State of the Art (Marjorie J. Clarke, Maarten de Kadt, Ph.D., and David Saphire), 1991, 275 pp., $47.

Garbage Management in Japan: Leading the Way (Allen Hershkowitz, Ph.D., and Eugene Salerni, Ph.D.), 1987, 152 pp., $15.

Garbage Burning: Lessons from Europe: Consensus and Controversy in Four European States (Allen Hershkowitz, Ph.D.), 1986, 64 pp., $9.95.

Improving Environmental Performance of MSW Incinerators (Marjorie J. Clarke), 1988, 82 pp., $15.

Energy and Air Quality

Paving the Way to Natural Gas Vehicles (James S. Cannon), 1993, 182 pp., $25.

Drive For Clean Air: Natural Gas and Methanol Vehicles (James S. Cannon), 1989, 252 pp., $65.

Other INFORM Publications

For a complete publications list, including materials on land and water conservation, and a quarterly newsletter, or for more information, call or write to INFORM.

Sales Information

Payment

Payment, including shipping and handling charges, must be in US funds drawn on a US bank and must accompany all orders. Please make checks payable to INFORM and mail to:

INFORM, Inc.
381 Park Avenue South
New York, NY 10016-8806

Please include a street address; UPS cannot deliver to a box number.

Shipping Fees

To order in the US, please send a check that includes $3 for the first book and $1 for each additional book for shipping and handling charges. To order in Canada, add $5 for the first book and $3 for each additional book. For information on shipping rates for other countries, call (212) 689-4040.

Discount Policy

Booksellers: 20% on 1-4 copies of same title
                30% on 5 or more copies of same title

General bulk: 20% on 5 or more copies of same title

Public interest and
community groups: **Price:**
    Books under $10:   No discount
    Books $10-$25: $10
    Books $25 and up: $15

Returns

Booksellers may return books, if in saleable condition, for full credit or cash refund up to 6 months from date of invoice. Books must be returned prepaid and include a copy of the invoice or packing list showing invoice number, date, list price, and original discount.

Membership

Individuals provide an important source of support to INFORM and receive the following benefits:

Member ($25): A one-year subscription to *INFORM Reports*, INFORM's quarterly newsletter.

Friend ($50): A one-year subscription to *INFORM Reports*, and advance notice of new publications.

Contributor ($100): Friend's benefits, plus a 10% discount on new INFORM studies.

Supporter ($250): Friend's benefits, plus a 20% discount on new INFORM studies.

Donor ($500): Friend's benefits, plus a 30% discount on new INFORM studies.

Associate ($1000): Friend's benefits, plus a complimentary copy of new INFORM studies.

Benefactor ($5000): Friend's benefits, plus a complimentary copy of new INFORM studies.

All contributions are tax-deductible.

INFORM's Board of Directors